BATTLE OF FRANKLIN

TENNESSEE

NOVEMBER 30, 1864

𝔄 𝔐onograph

By JACOB D. COX
Late Major-General Commanding Twenty-third Army Corps

WITH MAPS

NEW YORK
CHARLES SCRIBNER'S SONS
1897

Copyright, 1897,
BY CHARLES SCRIBNER'S SONS.

University Press:
JOHN WILSON AND SON, CAMBRIDGE, U.S.A.

This scarce antiquarian book is included in our special *Legacy Reprint Series*. In the interest of creating a more extensive selection of rare historical book reprints, we have chosen to reproduce this title even though it may possibly have occasional imperfections such as missing and blurred pages, missing text, poor pictures, markings, dark backgrounds and other reproduction issues beyond our control. Because this work is culturally important, we have made it available as a part of our commitment to protecting, preserving and promoting the world's literature.

CONTENTS

CHAPTER I

INTRODUCTORY 1

Results of the Atlanta Campaign — Hood's Movement on Sherman's Communications — Jefferson Davis's Relations to it — Beauregard's — Hood's March across Alabama — Delay at Tuscumbia — Sherman's Decision to March to the Sea — Thomas left in Tennessee — Schofield joins him — Strength of contending Forces — Problem of Concentration — Schofield at Pulaski — Hood's Advance — Schofield's Retreat to Franklin — Offer of Battle — Confederate Disaster — Tactical Problems — Comparisons — Erroneous Accounts — Official Records of the War — Schofield as Commander.

CHAPTER II

FROM COLUMBIA TO FRANKLIN 21

The Line of Duck River — Thomas urgent that Hood should be held back — Positions of the Armies — Hood begins the Flank Movement — Telegraphic Correspondence — How not to do it — Division of Confederate Forces — Schofield's Cavalry lose Communication with him — Combat at Spring Hill — Midnight March to Franklin.

CHAPTER III

TAKING POSITION AT FRANKLIN 37

Arrival at Franklin — No Bridge or Pontoons — Hood to be held back — Schofield's Oral Directions — His Correspondence with Thomas — Delay in Arrival of Reinforcements at Nashville — Can you hold Hood back three Days? — Orders to continue Retreat — The Position at Franklin — The Carter House — The Town and the River — The Field in Front — The Defensive Line — Repairing Bridges — Twenty-third Corps Positions — Reilly's Division — Ruger's Division — The Works on Carter Hill — Retrenchment across Turnpike — Kimball's Division.

CHAPTER IV

THE REAR GUARD — WAGNER'S DIVISION 64

Guarding the Trains — Collecting the Stragglers — Halt on Winstead Hill — Conditional Orders — Skirmishing with Forrest — Hood's Infantry press close — Will he turn the Position or attack ? — Wagner withdraws Lane's Brigade to Privet Knob — Conrad's farther in Rear — Opdycke's within our Works — Wagner's Message to Stanley — His Colloquy with Opdycke — His Orders to Lane and Conrad to Fight — The Sergeants to fix Bayonets — Ruger's Preparations — Confederates forming for the Attack.

CHAPTER V

THE CONFEDERATE ARRAY 83

Hood at Spring Hill — Discovers Schofield's Escape — Cavalry in Pursuit — Infantry hastening after — He decides to Assault — His Cavalry Positions — The Infantry — Deployment and Formation — Stewart's Corps on Hood's Right — Cheatham's in Centre and Left — Part of Lee's in Reserve — Artillery in Intervals — Chalmers's Cavalry on extreme Left — Hood's Headquarters.

CHAPTER VI

THE ASSAULT ON WAGNER'S OUTPOST 91

View from the Knoll on our Left — Skirmishing in Front — The Outpost trying to intrench — Confederate Advance — Colonel Capers's Description — Artillery opens on both Sides — Surgeon Hill's View from Fort Granger — Orders sent along our Line — To Opdycke in Reserve — Retreat of the Outpost — My Ride to the Centre — Momentary Break there — Reilly's Rally — Opdycke's Rush forward — Strickland's Rally — Meeting Stanley — The Din of Battle — Stanley wounded.

CHAPTER VII

THE FIRST FIGHT AT THE CENTRE 102

Hood's Advance retarded by the Outpost — His Right Wing farthest forward — My Staff at the Carter House — Wagner also there —

Messages from the Outpost — Wagner's Replies — Marshall's Guns come in — Disorganized Retreat of the Outpost — Wagner's Efforts to rally — Swept along to the Town — Cannon in the Enemy's Hands — But soon retaken — Fight over the Batteries — Reilly's Second Line charges — Fight at the Cotton-Gin — Destruction of Confederates — Heroism of their Officers — Reilly's Report — Opdycke's Formation for the Charge — Position of his Regiments — Of Strickland's — Focus of the Fight — Two Lines on Carter Hill — Turn of the Tide.

CHAPTER VIII

THE FIGHT OF OUR LEFT WING 121

Advance of Stewart's Corps — The narrowing Field — Thorny Hedges — Changes in the Array — Walthall's Report — Loring's Division strikes Stiles — Fighting in the Railway Cut — Batteries at Close Range — Loring Repulsed — The Attack on Casement — General Adams's Death — Our Line successfully held.

CHAPTER IX

THE FIGHT OF OUR RIGHT WING 130

Cheatham's Corps — Convergent Attack of Cleburne and Brown — Line of Lane's Retreat — Moore's Brigade Front unmasked — Our Artillery Cross-fire — Advance of Bate's Division — Battery at the Bostick Place — Close Quarters at Moore's Centre — Help from Kimball — Chalmers's Cavalry attack Kimball — Infantry attack his Left — Cavalry his Centre and Right — Confederates Repulsed.

CHAPTER X

THE SITUATION AT SUNSET 141

Determined Fighting at the Centre — Examination of Strickland's Line — Enemy holding outside of his Works — The Second Line — Relative Position of Opdycke and Strickland — Orders to the latter — Visit to Ruger — Wagner reorganizing — Visit to extreme Left — Reinforcements for the Centre.

CHAPTER XI

FROM THE CONFEDERATE STANDPOINT 148

Multiplication of Lines of Attack — How caused — Walthall's Description of the Assault — In the Abattis — Repulsed in Confusion — The Ditch at the Cotton-gin — Brown's Attack on Right Centre — Bate overlaps him — Johnson's Attack after Dark — Hood's Description — S. D. Lee's — Colonel Capers's — Capture of Gordon — His Account of the Charge — Cleburne falls.

CHAPTER XII

THE BATTLE AFTER DARK 160

In the Locust Grove — The Two Lines — The 112th Illinois — Reilly's Detachment — Sweeping the Ditches — Captain Cunningham's Story — The Sergeant Major's — General Strahl's Death — Hood's Reserves — Rallying on them — Later Alarms — Orders to Wood's Division — Preparations for Withdrawal.

CHAPTER XIII

WILSON'S CAVALRY ENGAGEMENT 172

Morning Positions — Covering both Flanks of the Army — Forrest's Advance — The Fords of the Harpeth — Confederates cross at Hughes's Ford — Wilson attacks — Sharp Combat — Enemy retreat across the River — Covering the March to Nashville.

CHAPTER XIV

OUR WITHDRAWAL 180

The Medical Department — Field Hospitals — Ambulance Train — Sick and Wounded sent to Nashville — Work of Surgical Corps during the Battle — Artillery gradually withdrawn — Arrangement of Skirmish Lines — Movement of the Infantry — Orders as to Kimball's and Wagner's March — Misunderstanding — March of Ruger, Opdycke, and Reilly — A burning Building — The Field in Front — The March to Nashville.

CHAPTER XV

FRANKLIN AFTER THE BATTLE 194

Hood's Midnight Order — Condition of his Army — Discussion of Probabilities — Confederates move by the Flank — Experiences of the Carter Family — Colonel Carter's Story — Defensive Armor — Refuge in the Cellar — The terrible Night — Captain Carter's Fate — Private Gist's Adventures — General Cooper's Retreat.

CHAPTER XVI

RESULTS AND LESSONS 207

Sources of Statistical Knowledge — Hood's Forces before the Battle — Schofield's — Numbers actually engaged — Hood's Casualties — Loss of Officers — Schofield's Losses — Analysis of them — The Problem of Attack and Defence — Fire Discipline.

CHAPTER XVII

DISCUSSION OF WAGNER'S CONDUCT 220

Natural Rise of Controversies — Corps Feeling — Good Comradeship — Wagner's Personal Situation — Disposition to befriend him — Criticisms by his Subordinates — Efforts to allay the Irritation — Correspondence — Preliminary Reports — Conrad's Report — Wagner retired from the Division — Leaves the Army.

CHAPTER XVIII

DOUBLE BREASTWORKS ON CARTER HILL 233

The Two Lines at our Right Centre — Conflicting Memory of Eye-witnesses — Confederate Testimony — Solution of the Matter — Bullet Marks on Brick Smoke-house — Summary of the Evidence.

CHAPTER XIX

THE RALLYING OF THE OUTPOST BRIGADES 243

Value of Detailed Reports — Landmarks — Lines of Retreat from the Outpost — Crowding toward the Centre — Significant Omissions — Lists of Missing — What they Teach — Reports Compared — Incidents relating to Captured Flags — Statements of various Officers — Conclusions from the Facts.

CHAPTER XX

An Unexpected Controversy 258

Colonel Stone's Paper in Century War Book — General Stanley's Criticism — A Violent Attack — Earlier Correspondence — Nine Points — Two Corrections — Basis of a Historical Narrative.

CHAPTER XXI

Controverted Points 266

Work assigned the Twenty-third Corps — Detachment of Fourth Corps Batteries — Orders to Ruger and Kimball — Detachments often Necessary — Articles of War and Regulations — Questions of Command — Stanley on North Side of River — His Ride to the Front — Soon Wounded — He Retires — Statements of Officers — At the Field Hospital — At Schofield's Headquarters — Summary — Official Reports — Analysis of Stanley's — Contemporaneous Records — Conclusion.

APPENDICES

A. General Schofield's Report 305
B. General Cox's Report 311
C. General Wood's Report 323
D. General Stanley's Report 327
E. Colonel Dow's Statement 332
F. Colonel Cox's Statements 336

INDEX 341

BOOKS OF FREQUENT REFERENCE

Their Full Titles

Official Records of the Union and Confederate Armies. Series I. Referred to by the initials O. R., with the numbers of the original volumes in parts.

Battles and Leaders of the Civil War. 4 vols. The Century Co. 1884–88. Referred to as Century War Book.

Sketches of War History. Papers published by the Ohio Commandery of the Loyal Legion. Referred to as Ohio L. L. Papers.

History of the Army of the Cumberland. By Thomas B. Van Horne, Chaplain U. S. A. 2 vols. 8vo, and Atlas. Cincinnati: Robert Clarke & Co. 1875. Referred to as Van Horne's Army of the Cumberland.

The Life of Major General George H. Thomas. By Thomas B. Van Horne, U. S. A. New York: Charles Scribner's Sons. 1882. Referred to as Van Horne's Life of Thomas.

Narrative of Military Operations directed, during the late War between the States, by Joseph E. Johnston, General C. S. A. New York: Appleton & Co. 1874. Referred to as Johnston's Narrative.

Rise and Fall of the Confederate Government. By Jefferson Davis. New York: Appleton & Co. 1881. Referred to as Davis's Rise and Fall, etc.

Advance and Retreat: Personal Experiences in the United States and Confederate States Armies. By J. B. Hood, Lieut. General Confederate Army. New Orleans. 1880. Referred to as Hood's Advance and Retreat.

Campaigns of the Civil War: Atlanta. By Jacob D. Cox, late Major General Commanding Twenty-third Army Corps. New York: Charles Scribner's Sons. 1881. Referred to as Atlanta.

Campaigns of the Civil War: The March to the Sea, Franklin and Nashville. By Jacob D. Cox, etc. New York: Charles Scribner's Sons. 1882. Referred to as Franklin and Nashville.

Campaigns of the Civil War: Statistical Record of the Armies of the United States. By Frederick Phisterer, late Captain U. S. A. New York: Charles Scribner's Sons. 1883. Referred to as Phisterer's Statistical Record.

LIST OF MAPS

1. Map of Middle Tennessee and Northern Alabama.
   To face Chapter I.
2. Sketch Map of Carter Hill, from Survey by Colonel M. B. Carter.
   To face Page 43.
3. Map of the Battle-field of Franklin. By Major W. J. Twining, U. S. A., Chief Engineer Army of the Ohio. The original of this Map is in the Atlas accompanying the Official Records, Plate CXXXV., C, No. 5.
   To face Page 45.
4. Map of the Battle-field of Franklin. By Major W. F. Foster, Engineer of Stewart's Corps, Confederate Army. The original of this Map is in the Atlas accompanying the Official Records, Plate LXXIII., No. 3.
   To face Page 83.

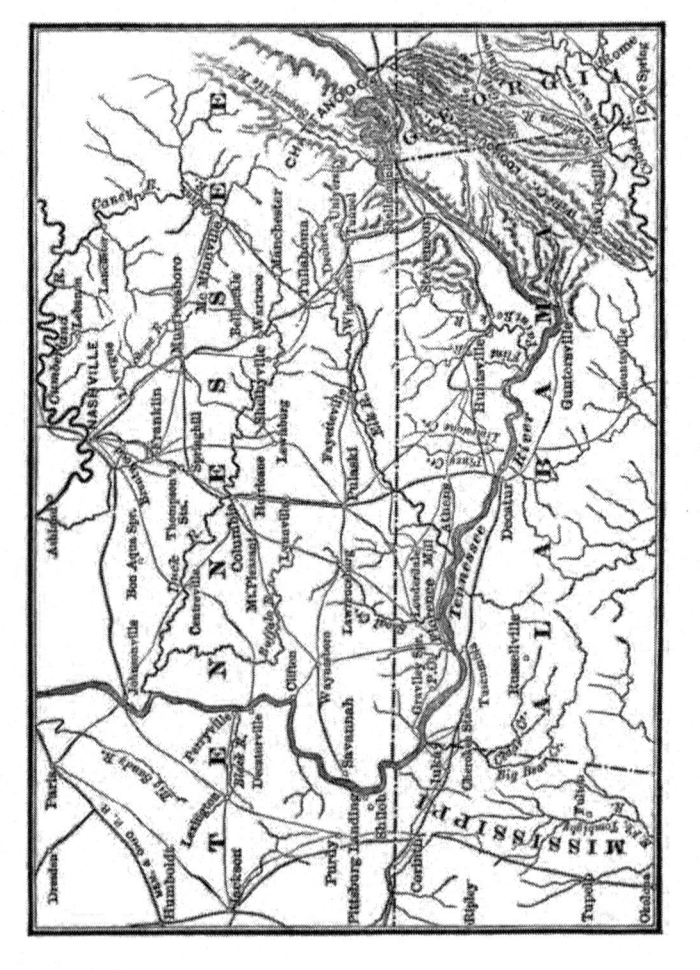

MIDDLE TENNESSEE AND NORTHERN ALABAMA.

THE BATTLE OF FRANKLIN

CHAPTER I

INTRODUCTORY

Results of the Atlanta Campaign — Hood's Movement on Sherman's Communications — Jefferson Davis's Relations to it — Beauregard's — Hood's March across Alabama — Delay at Tuscumbia — Sherman's Decision to March to the Sea — Thomas left in Tennessee — Schofield joins him — Strength of contending Forces — Problem of Concentration — Schofield at Pulaski — Hood's Advance — Schofield's Retreat to Franklin — Offer of Battle — Confederate Disaster — Tactical Problems — Comparisons — Erroneous Accounts — Official Records of the War — Schofield as Commander.

THAT the battle of Franklin was a hard fought and bloody combat is now generally known; but this of itself would not warrant a monograph giving with some fulness of detail the progress and the incidents of the fight. There must be a limit to the minuteness of history even in the most important events, and in military history no less than in the narration of civil affairs. When, however, a battle proves to be a turning point in a decisive campaign, — when it marks the "beginning of the end" in such a contest as our civil war, — when it justifies the strategy of such a leader as Sherman in his division of his forces in Georgia and making the March to the Sea, — when in addition to this the combat may be fairly said to be a crucial experiment in the problem of attack and defence of fieldworks in an open country, — we can hardly place a limit to the desirability of detailed knowledge. Everything which helps to a complete understanding is then made welcome. Even those things which at first blush may seem trivial are not so if they aid us in comprehending how men live, and act, and think, and fight, and die, on such a stubbornly contested field.

In all these respects the battle of Franklin was an important one; but besides these claims upon historical attention, it was for a long time greatly misunderstood, and controversies of all sorts grew out of it and the campaign of which it was a part. For thirty years the author has been frequently urged by his comrades to accept the task of writing the story of the battle, and when he had prepared a brief history of the campaign,[1] the desire for a much fuller account of the battle itself did not appear to be satisfied. He made conditional promises that, if he should live to see the completion of the great work undertaken by the government in printing the "Official Records of the Union and Confederate Armies," he would accept the duty. The last volumes of the principal series of those records are passing through the press, and the following pages are written in the effort to redeem his promise.

Soon after the fall of Atlanta, Mr. Davis, the Confederate President, visited General Hood in his Georgia camp for a conference upon the further prosecution of the campaign. Hood's energy was indomitable, and he still believed in aggressive strategy. In this he had the sympathy of Davis, who had constantly shown his dissatisfaction with Johnston's defensive policy, and who was ready to believe Hood's

[1] The March to the Sea: Franklin and Nashville. Charles Scribner's Sons, 1882.

over-sanguine assertions that his army had not suffered in its *morale* by the bloody and unsuccessful engagements around Atlanta or by the loss of that important strategic point. Hood's proposal to turn Sherman's position by a somewhat wide detour to the west of Atlanta, and to carry the war again into Northern Georgia, or even into Tennessee, met with Mr. Davis's complete approval, and in a speech to the troops he was so far carried away by the desire to stimulate their confidence and their courage that he imprudently announced the general plan of campaign.

The speech got into the newspapers, and came to Sherman's ears. It need hardly be said that, with his characteristic alertness of mind and quickness of decision, he set himself the task of defeating his enemies' expectation. He would not give them the appearance even of success, and at the last moment, before breaking his communications with the North, and starting on his March to the Sea, he used Davis's defiantly proclaimed purpose to overcome the last of Grant's doubts. "The whole burden of his song," he wrote to the General in Chief, "consisted in the statement that Sherman's communications must be broken, and his army destroyed. Now it is a well settled principle that, if we can prevent his succeeding in his threat, we defeat him, and derive all the moral advantages of a victory."[1] Sherman gave Hood full credit "for his skilful and rapid lodgment made on the railroad,"[2] but pointed out that he had preserved his own communications by his rapid chase of the enemy, and had taken from Hood the advan-

[1] Sherman to Grant, November 6, 1864, O. R., xxxix. part iii. p. 659.
[2] *Ibid.*

tage of the movement, by forcing him into the long march in retreat across northern Alabama into Mississippi. "To have dogged him far over into Mississippi, trusting to some happy accident to bring him to bay and to battle," though much might be said in favor of it, "would be," he concluded, to "play into his hands by being drawn or decoyed too far away from our original line of advance."[1] So he cut the knot, and gave orders for the concentration of his own column at Atlanta, and for the march through Georgia.

The game, at this stage, was a magnificent one. Hood's plan proved a military character and mind of no ordinary powers, and he had carried it out with a celerity and vigor that were in complete keeping with it. The probabilities were all in favor of his success. The odds were a hundred to one that his march to Tuscumbia, in the northwest corner of Alabama, would be met by Sherman's transfer by railroad to meet him at Columbia or Pulaski, in southern Tennessee. Grant's dispatches show that he would have continued the pursuit, and this alone is a sufficient tribute to Hood's ability.[2] Sherman met him with a still greater audacity than his own, and with a success followed by far reaching consequences, which precipitated the fate of the Confederacy.

Under pressure from influential men in the Confederate Congress, Mr. Davis had assigned General Beauregard to the command of the two departments which had been under Generals Hood and Taylor, but with limitations which implied that he should

---

[1] Sherman to Grant, November 6, 1864, O. R., xxxix. part iii. p. 659.
[2] *Id.*, p. 576.

not assume active command of either army in the field unless an exigency made it necessary. Hood was disposed to make the most of his independence, and Beauregard limited himself generally to advice and suggestions, the larger plan being supposed to have been settled between Davis and Hood at the beginning of the new campaign.[1] Beauregard's relations to the active army remained merely nominal, though he gradually became more pressing in his urgency that Hood should resume vigorous offensive activity. The latter found the railways near Tuscumbia badly out of order, and thought it necessary to accumulate twenty days' supplies before crossing the Tennessee River.[2] General Wheeler, the cavalry commander, who had been left near Georgia, had sent him accurate information of the movements of Sherman's different corps, and of the statements of prisoners that Sherman was expecting to march to Savannah; but he preferred to interpret the news in accordance with his wishes, and on the 4th of November reported to Richmond that Sherman seemed to be concentrating his army at Huntsville and Decatur in Alabama.[3] When he finally became convinced that Sherman in person, with a veteran army, was marching toward the Atlantic coast, he was torn by conflicting motives as to his own course, and it required the peremptory orders of Beauregard to put an end to his delay. But it was now the last week in November; nearly a month had been given Sherman and Thomas to complete the arrangements

[1] See dispatches between Davis and Beauregard, O. R., xxxix. part iii. pp. 782, 785, 870, 874, 879.

[2] Hood to Taylor, *Id.*, p. 868; to Seddon, Secretary of War, p. 870; to Railway Officials and Quartermaster, p. 871; to Beauregard, p. 888.

[3] Wheeler to Hood, O. R., xxxix. part iii. pp. 859, 878. Hood to War Department, *Id.*, p. 888.

for a double campaign, and Sherman's start was too long a one, both in time and in distance, to give the Confederates any hope of overtaking him. To force the fighting with Thomas was the only course left, and this was what Beauregard ordered.[1]

The contrast between the boldness and vigor of Hood's October campaign and his delay after reaching Tuscumbia points to a consciousness on his part that he had been outgeneralled, though he kept up a show of satisfaction at Sherman's division of his army, followed by the wide separation of its parts. He even wrote to Davis, that, should Sherman "move two or three corps south from Atlanta," he thought it would be "the best thing that could happen for the general good" of the Confederacy.[2] But Davis reminded him that this implied that prompt advantage should be taken of the dispersion to beat his adversary in detail.[3] The troops left under General Thomas were necessarily scattered, and for Hood to give him time to concentrate was to imperil the campaign. His true policy was, plainly, to risk everything on the quickest and strongest advance against Thomas, living upon the country, or trusting to success to give him possession of some of the national depots of supply. He was at Tuscumbia on the 30th of October, and next day reported that he was in possession of Florence (on the north bank of the Tennessee), and was laying a pontoon bridge.[4] He did not begin his advance northward till the 21st of November.[5] Knowing, as we do, the energy of his

---

[1] Beauregard to Hood, November 17, O. R., xlv. part i. p. 1215. Hood reported that he had only seven of the twenty days' needed supplies on hand. *Ibid.*

[2] O. R., xxxix. part iii. p. 891.   [3] *Id*, p. 896.

[4] *Id.*, p. 870.   [5] O. R., xlv. part i. p. 1236.

character, it is incredible that he should have waited so long unless he were hoping that something would happen to make it feasible to follow Sherman, by showing that the latter was moving upon Selma or Mobile instead of Savannah.

In arranging his plans for the March to the Sea, Sherman had from the first determined to put Thomas in general command of all that was left behind, and which lay east of the Mississippi River. For himself, he did not mean to make a subordinate operation with a detachment, but to move an army which would be fit to make the campaign of the Carolinas after he should reach Savannah, and which could cope even with Lee's army if it should break away from Richmond. Definite plans beyond Savannah were postponed till he should be successful in reaching the coast, but the contingency of a march northward was prominent in his mind from the beginning. As he must move from Atlanta without lines of supply or communication, he must strip his army of every unnecessary weight or encumbrance, and must look for no help or reinforcement on the way. His column must necessarily diminish by the attrition of the campaign, and would be strongest at the start. Thomas's, on the other hand, would be weakest at the start, but would be rapidly increased by new troops which were moving to the front, by return of the convalescents of the whole army, and by detachments from the departments west of the Mississippi. The principal of these last was a portion of the Sixteenth Corps, supposed to be ten thousand strong, under General A. J. Smith, who was already under orders to proceed from Missouri to Tennessee.[1]

[1] O. R., xxxix. part iii. pp. 477, 494, 595.

Sherman had at first planned to send back the Fourth Corps, and to take the Twenty-third Corps with him, leaving General Schofield to support General Thomas with the troops of the Department of the Ohio which were in Kentucky and East Tennessee. But when Schofield returned from absence at the North, and rejoined the army at Gaylesville, Alabama, on October 22d, he was very unwilling to part with the corps. It was, besides, much reduced by the casualties of the year's campaigns, and greatly needed to be recruited. When, then, we got news that Hood had appeared at Decatur, and, being repulsed there, had marched to Tuscumbia, Schofield suggested that, if the corps were sent into Tennessee, it could be recruited, a division of new troops could be added to it, and he himself could take active field service under Thomas. Before that time it had seemed probable that Hood would follow Sherman southward with at least the greater part of his army; but his committal to a movement in force into middle Tennessee made it apparent that Thomas might need reinforcement before troops from distant positions could come to him. Thomas was fearful that troops from Missouri could not reach him soon enough, but said that if he had Schofield he should feel perfectly safe. On the 28th of October, therefore, Sherman announced his intention of sending the Twenty-third Corps, as well as the Fourth, to General Thomas, and to take with him the remainder of his army, thus reduced to about fifty thousand men. It turned out that the hurry of absentees to rejoin their commands increased his numbers somewhat, before communications were broken.[1]

[1] O. R., xxxix. part iii. Sherman to Halleck, October 27, p. 461; Schofield to Sherman, p. 468; Sherman to Halleck, October 28, p. 476;

Introductory 9

The effective force under General Thomas, in middle and southern Tennessee, was sixty-five thousand five hundred, officers and men "present for duty equipped," which was the official phrase indicating complete readiness for active service. The aggregate present was some twenty thousand more.[1] These figures do not include the troops in Schofield's Department of the Ohio in East Tennessee and Kentucky, nor those of the Military Division between the Tennessee and Mississippi Rivers, all of which were subject to Thomas's orders. Hood's army in the field numbered forty-two or forty-three thousand men of all arms, and had been very closely estimated by General Sherman.[2] Thomas's problem, therefore, was chiefly one of concentration in time to meet Hood's advance, and the delay of the latter, whether necessary or not, was more than could have been hoped for. As early as October 29th, Sherman, in promising to send Schofield back, had urged Thomas to "break up all minor posts, and get about Columbia as big an army as you can, and go at him."[3] Again, two days later, he reiterated, "You must unite all your men into one army, and abandon all minor points, if you expect to defeat Hood."[4] In pursuance of this policy, Thomas ordered Schofield with the Twenty-third Corps to join Stanley and the Fourth Corps at Pulaski, on the railroad between

Sherman to Beckwith, p. 477; Thomas to Halleck, p. 582; Sherman to Thomas, pp. 484, 497, 499, 514.

[1] Tri-monthly Official Return of General Thomas for November 20, 1864, which was the first after the severance from Sherman. The effectives were increased by 9,700 in the next ten days. O. R., xlv. part i. pp. 52, 53, 54. As to Hood's forces, see chap. xxi., *post*.

[2] Sherman to Grant, O. R., xxxix. part iii. p. 576; to Thomas, *Id.*, p. 599.

[3] *Id.*, p. 498. [4] *Id.*, p. 535.

Columbia and Decatur, though it was necessary to send one brigade temporarily to Johnsonville, on the Tennessee River west of Nashville, to meet a raid that Forrest was making in West Tennessee by way of diversion in favor of Hood.

General Schofield reached Pulaski on November 13th,[1] and assumed command of the two corps of infantry, and of the cavalry corps under General James H. Wilson, as the latter gradually assembled, and the army in the field consisted of these organizations until the general concentration at Nashville on December 1st, after the battle of Franklin. Each of these corps was reinforced by some new troops, and some small detachments of older ones, but their general organization remained. A question of rank between General Schofield and General Stanley was before the War Department, but General Thomas had not heard of its decision in favor of Schofield when he assigned him to command in the field. He informed General Halleck of what he had done, and gave his reasons for hoping that his action would be sustained, as was in fact done.[2] Sherman advised that the two veteran corps should be filled up to twenty-five thousand each, and that Thomas should take the field in person when the enemy should advance. Thomas replied that such was his purpose.[3] Both the corps, however, had only the force intended for one, in the period of the campaign we are now considering, and at Pulaski only one division of the Twenty-third Corps was present.

[1] O. R., xxxix. part iii. p. 768; *Id.*, xlv. part i. pp. 885, 886.
[2] Thomas to Halleck, O. R., xxxix. part iii. p. 666; to Schofield, p. 691. See also Thomas's order, *Id.*, p. 638; Grant to Halleck, *Id.*, p. 684; Halleck to Sherman, *Id.*, p. 64.
[3] *Id.*, p. 685.

Hood moved his infantry forward from his bridge at Florence on the 21st and 22d of November,[1] marching upon Columbia by way of Lawrenceburg and Mount Pleasant, turning Schofield's position at Pulaski upon the Nashville and Decatur Railroad. Schofield withdrew his little army in echelon, by way of Lynnville and Hurricane, keeping just ahead of Hood in the convergent movements upon Columbia, and reaching the latter place on the 24th, barely in time to prevent Hood from anticipating him. Our army remained there till the 27th, when the enemy's operations to turn the position by the other flank (our left) forced our retirement to the north bank of Duck River. There, the more detailed account of events preceding the battle of Franklin will begin.[2]

At Franklin, on the 30th of November, Hood was pressing so close upon our columns that it became necessary to choose between abandoning our wagon trains and resisting him on the south side of the Harpeth River. Schofield resolved to take the risk of fighting although the stream was at our backs, and the result fully justified him. The repulse of the Confederates was so destructive, that, though Thomas thought it wise to withdraw Schofield to Nashville and make his concentration there where A. J. Smith's troops had just arrived, Hood's advance from the Harpeth was the merest bravado, and was naturally followed by the final ruin of his army. No intelligent officer on either side was ignorant of the fact that the heart of the Confederate army was broken, and the character of the fighting was, from that day, in marked contrast with what it had been before.

[1] O. R., xlv. part i. p. 1236.
[2] Chap. ii., *post*.

Anticipating the judgment which history must give, ranking the battle of Franklin as the crisis in the campaign and the veritable "beginning of the end," Hood, in his report of December 11th, treated it as a necessary result of the failure of his subordinates on the 29th of November, at Spring Hill, to obey his orders with energy. There, he said, "was lost the opportunity for striking the enemy for which we had labored so long, the best which this campaign has offered, and one of the best afforded us during the war."[1] A critical examination does not justify this effort to shift the responsibility, but it remains a very significant recognition of the fact that Franklin sealed the fate of the campaign. Jefferson Davis clearly, though in cautious phrase, bears witness to the same truth when, in his "Rise and Fall of the Confederate Government," he says of the battle, that it was "one of the bloodiest of the war, whose results materially affected the future."[2] Davis sympathized with and defended the brilliant aggressive strategy of the campaign, though he limited his own advice to the operations within the State of Georgia, and the confines of northeastern Alabama. He had removed Johnston from command because of his persistent Fabian policy, but was forced by the results to throw doubts upon the wisdom of the attack at Franklin. Beauregard, Hood's immediate superior, in his general report of April 15th, added his testimony by saying, "It is clear to my mind that, after the great loss and waste of life at Franklin, the army was in no condition to make a successful attack on Nashville."[3] He would rather

---

[1] O. R., xlv. part i. p. 657.
[2] Rise and Fall of the Confederate Government, vol. ii. p. 575.
[3] O. R., xlv. part i. p. 651.

have moved on Murfreesboro, and then sought winter quarters behind the Duck or Tennessee River, detaching a force for the protection of South Carolina. His tribute to the character and courage of the troops, all who had personal knowledge of the campaign will unite in. "Untoward and calamitous as were the issues of this campaign," he said, "never in the course of this war have the best qualities of our soldiery been more conspicuously shown; never more enthusiasm evinced than when our troops once more crossed the Tennessee River; never greater gallantry than that which was so general at Franklin; and never higher fortitude and uncomplaining devotion to duty than were displayed on the retreat from Nashville to Tupelo."[1] General Johnston supports, in substance, both parts of Beauregard's statement, though his words are strongly tinged with sarcasm at the results which followed his own removal from command. Of the officers and men of that army, he says, "Their courage and discipline were unsubdued by the slaughter to which they were recklessly offered in the four attacks on the Federal army near Atlanta as they proved in the useless butchery at Franklin, and survived the rout and disorganization at Nashville as they proved at Bentonville."[2]

I need not go beyond these Confederate authorities of the first rank for support of my assertion that this battle was the turning point in a great campaign, and fully justified Sherman's plan. Its character as a crucial experiment in the problem of attack of field

---

[1] O. R., xlv. part i. p. 651.
[2] Narrative of Military Operations, p. 365. The battle of Bentonville, North Carolina, was the last engagement between Sherman and the Confederates, and Johnston had been restored to his command. For the four attacks by Hood near Atlanta, see my volume "Atlanta" (Scribner's Campaigns of the Civil War), chap. xii.

fortifications will appear in the progress of my narrative. Hood bore with patience the penalty of failure, but justice requires the clear acknowledgment that his faith in the attack has been, and perhaps still is, the prevalent military creed in Continental Europe. His tactics of assault in deployed line with supports are not far from the approved methods supposed to be developed by late wars. It would be hard to match in dash or in perseverance the veteran Confederate battalions of 1864; it would be impossible to surpass the leadership of the officers who headed the charges upon the field. If theory is worth testing at all, no real soldier will be hard on Hood for testing it. It was, without doubt, our general conclusion in 1864–65, that the strategic initiative and tactical defence had the best of it. But the race characteristics and the educated qualities of men make such a difference that it would be rash to conclude that American experience is conclusive for Germans or Frenchmen. To make a rapid march, then halt and construct efficient cover whilst a vigorous enemy is approaching, requires an intelligence, alert handiness, and coolness in the rank and file and in the subordinate officers which cannot always and everywhere be expected. The ability to await an attack is also a very variable quality. In bayonet charges, it is notorious that one side or the other usually breaks before bayonets are actually crossed. With material like our frontiersmen, among whom a half-dozen cool and determined soldiers and scouts, burrowing in a "buffalo wallow" upon the plains, will "stand off" a hundred Indians, almost anything can be done.[1] The

---

[1] General Nelson A. Miles's Personal Recollections, p. 173.

reader must judge how far the personal element counted at Franklin on both sides.

The comparative smallness of the opposing armies is likely to lead to an under estimate of the desperate character of the fighting. The analysis of the forces engaged in the actual attack and defence will come later.[1] It is enough now to note the fact that Hood had more men killed at Franklin than died on one side in some of the great conflicts of the war where three, four, or even five times as many men were engaged. His killed were more than Grant's at Shiloh, McClellan's in the Seven-days' battle, Burnside's at Fredericksburg, Rosecrans's at Stone's River or at Chickamauga, Hooker's at Chancellorsville, and almost as many as Grant's at Cold Harbor. The concentration in time, in those few hours of a winter afternoon and evening, makes the comparison still more telling.[2] It may be properly said, therefore, that the intrinsic importance of the battle, from the purely military point of view, warrants a full study of its history in all its features.

The early treatment of the subject was peculiarly unsatisfactory to the great majority of the officers and men who participated in it on either side. This was in great part due to the fact that the period following the war was one in which military writings took naturally the form of personal memoirs or

---

[1] See chap. xvi., *post*.

[2] See tables in Phisterer's Statistical Record of the Armies of the United States (Scribner's Campaigns of the Civil War), pp. 213–218. The comparisons are necessarily made with losses on the National side, because no analyzed tables of the Confederate losses exist. The number of Hood's killed at Franklin is that given in General Thomas's official report, i. e. 1,750. See chap. xvi., *post*. Colonel Maurice gives the number of those actually killed at Waterloo, out of the British army of 43,000, as 1,759. Maurice, War, p. 83.

histories of particular organizations, such as regiments, brigades, divisions, and corps. In the narratives of the experience of a single organization, it was not to be expected that a symmetrical history of a campaign should be given, and there would be little ground for complaint that other organizations should receive scant attention, even when all were busily participating in the events described. In regard to the larger organizations, however, such as corps and wings of a larger army, it was almost inevitable that such a mode of writing should be misleading. The whole history of a campaign or battle would seem to be given, and the reader would naturally assume that he had the whole, unless he were distinctly told that the narrative was confined to a part of the army.

In the campaign of November, 1864, against Hood, the two organizations of the Army of the Cumberland and the Army of the Ohio were each represented by a corps. Only those quite familiar with the army know this, and when Chaplain Van Horne wrote his history of the Army of the Cumberland, including this campaign, it was commonly supposed to cover the whole field and the services of both corps referred to. The book was an important one among the earlier historical publications concerning the Civil War, and was largely quoted and followed by general historians of the period. As to the battle of Franklin, the troops which occupied during the whole battle five sixths of the line, a mile long, on which Hood made his desperate infantry attacks, found themselves ignored. They read accounts of the battle in which they were not even mentioned. Time, of course, gradually corrected some of the errors and misapprehensions. In a second book, the "Life of

General Thomas," Van Horne materially modified his earlier treatment of the subject, and handled it in a more comprehensive way. Other noteworthy papers appeared, written by participants in the campaign or by eminent journalists and literary men who personally visited the field.[1] A much more correct idea of the campaign and of the battle was the result of this free discussion, although for a time each new publication seemed to start new questions for debate. It was entirely natural that my own statements of fact, in my little volume already mentioned, should be questioned; for the brevity of treatment made it impossible to give the wide array of public documents and private evidence which supported my own recollection. As new points appeared, I have tried to obtain the well considered remembrance of surviving actors in the campaign, but the reader will notice how largely this had been done prior to the publication of my book in 1882. It will thus be seen that I did not speak at random, and that I used reasonable diligence in testing my own memory before making historical statements. Absolute freedom from error was, of course, improbable; but more thorough investigation has fully sustained the general accuracy of my narrative.

The publication of the Official Records begins a new epoch in the study of our Civil War. The

[1] Among these were a narration by Gen. Emerson Opdycke, New York Times, Sept. 10, 1882; F. H. Burr, in Philadelphia Press, March, 1883; Rev. Dr. H. M. Field, in New York Evangelist, and in his book "Bright Skies and Dark Shadows"; Prof. W. W. Gist, in Cedar Rapids, Iowa, Republican, September, 1883; Col. Henry Stone, in Century War Book; Capt. Thomas Speed and Capt. L. T. Scofield, in Loyal Legion papers of Ohio Commandery; Maj. D. W. Sanders (Confederate), in Southern Bivouac, 1885; Gen. B. F. Cheatham, in Louisville Courier Journal, December, 1881; General Hood's "Advance and Retreat," etc.

fortunate preservation of Confederate records and their careful collection by the commissioners must be a matter of congratulation amongst all lovers of truth, and of pride in the heart of every American. The notion of saving only the records of our own side was promptly discarded. The whole truth was wanted, and as part of the officially collected records of the Confederate government were lost in the closing scenes of the conflict, neither labor nor cost was spared to replace missing papers by authentic copies which had been privately preserved. The voluminous mass which fills the hundred bulky volumes may seem at first like an excess of material; but there can be no excess in such a matter. Formal reports may give distorted views of past events, but the current dispatches and correspondence from day to day and hour to hour expose all glozing of errors or false claims of victory. We cannot have too perfect a presentation of the details which make up the actual life of the scene, with the hopes and the purposes, the current knowledge and apprehension, the hesitation and the decisive act of every responsible participant in such great events. In this library of telltale evidence each officer of any prominence has accumulated the proofs of character as to accuracy, candor, judgment, modesty, and, when the question of the value of his statement or the weight of his opinion is raised, the historian will not have far to seek for the means of determining it. In the same cloud of witnesses will be found the means of settling the truth when formal statements are conflicting. Established facts will be fixed by the general current of evidence; and accord with these, or disagreement, will test the correctness of particular reports or claims. Bearing these things in mind, I have not spared pains in verifying my recollection by references to the Official Records, so that the reader who chooses to do so may test the statements of the text by the original sources of knowledge in this field.

A few matters have been subjects of discussion or of controversy to such an extent that their complete treatment would too much impede the current of continuous narrative. My method, therefore, will be to tell the story first, with, as nearly as may be, an equal fulness of detail in the parts, and then devote a few chapters to the more particular discussion of some controverted points, taking up each in turn.

When we fasten our attention upon the subordinate parts in such a battle as Franklin, we seem to lose sight of the most important of all, — the controlling will which directs the whole. This is especially true as to the national side, because the plan of defence was a simple one, and there were few opportunities for new orders after the battle was opened. It must not be forgotten that the decision as to each step in the movement from Pulaski to Franklin had rested with General Schofield as commandant in the field. This decision had to be made in what were sometimes very trying circumstances, as at Columbia and Spring Hill. It was his responsible duty to decide at Franklin whether he would place his forces, as they arrived, upon the north bank of the Harpeth River, running the great risk of the loss of his army trains, or give battle on the south side, in front of the town, with the great disadvantage of having the river at our back. When he had determined to fight, he had to fix the general plan, and give the orders which alloted to his subordinates their positions and their duty. He was always ready to listen to suggestions and opinions, and was most considerate in allowing all reasonable discretion in the performance of duty by those under him; but he fully assumed his own burden of responsibility, and decided promptly what his duty called upon him to decide. He selected his own position with thoughtful reference to the efficient performance of his own task, present or future, according to the actual situation or the probable sequence of affairs.

But it is not only in the foresight shown in planning and in issuing the larger orders that a commanding officer makes himself felt by his immediate subordinates. He can do much by keeping up frequent communication with them, even though no new directions are called for. His messages of encouragement, of inquiry, of suggestion in case of contingencies, in a hundred ways keep up the contact of mind with mind, and carry his own spirit and will to those who are executing his commands. In all these respects, General Schofield had a high ideal of duty, and fully performed it. If, then, the account of the battle of Franklin shall seem to the reader to be mostly made up of the personal actions of those who were upon the fighting line, from private soldiers upward, he may well be reminded that, from beginning to end, those whose duty it was to carry out the purposes of the commander found themselves in constant touch with him, and were conscious of the distinct influence of his courage, his confidence, and his full comprehension and command of the situation.

CHAPTER II

FROM COLUMBIA TO FRANKLIN

The Line of Duck River — Thomas urgent that Hood should be held back — Positions of the Armies — Hood begins the Flank Movement — Telegraphic Correspondence — How not to do it — Division of Confederate Forces — Schofield's Cavalry lose Communication with him — Combat at Spring Hill — Midnight March to Franklin.

On Monday, the 28th of November, 1864, the army under General Schofield was holding the line of the Duck River, opposite Columbia. Hood's advance from Florence, on the Tennessee, had been by way of Lawrenceburg upon Columbia, and had made our position at Pulaski untenable, for the route of the Confederates carried them far beyond our right flank. To gain time for the expected arrival of reinforcements from Missouri under General A. J. Smith was the eager wish of General Thomas, who was in supreme command, and his dispatches to General Schofield from Nashville, where his headquarters were, had constantly suggested the most dilatory strategy. Each day was of great value, and if he could hold Hood south of Columbia and the Duck River for a week, he believed that he could bring enough reinforcements to that point to make him willing to seek a decisive engagement with the Confederate army on the ground which had been pointed out to him by General Sherman as the proper place for concentration.[1] Whilst, therefore, he acquiesced in the retrograde steps which Schofield was obliged to take, it was with manifest reluctance. His assent was usually coupled with the condition that his subordinate should find the necessity still manifest after careful consideration. He indicated the steps in the approach of his reinforcements, and the little margin of time he needed to make him strong enough to resume the aggressive. The effect of this was that Schofield felt bound to take considerable risks, and to delay each step in retreat until (as he said in his dispatch of the day of the battle of Franklin) the slightest mistake on his part or failure of a subordinate might have proved disastrous.[2] From Thomas's point of view his conclusions were the natural ones, and his strong desire to hold back his adversary was most proper; we are now considering only its effect upon Schofield in making him stick to the positions he occupied more stoutly than his own judgment would have dictated. When he abandoned Columbia in the night of the 27th, he felt so strongly the disappointment it would be to Thomas that his dispatch announcing the fact was apologetic in tone.[3] He said that he regretted exceedingly the necessity, but believed it to be absolute, and would explain fully in time.

It is necessary to keep this phase of the situation very clearly in mind; for Schofield's critics have made the deliberateness of his retreat a favorite point for attack; overlooking the pressure upon him to make it still more so, and the indisputable fact that, had he

[1] For Sherman's directions on this point, see O. R., xxxix. part iii. pp. 498, 535, *et seq.*; for Thomas's correspondence with Schofield, *id.*, xlv. part i. pp. 896–1170.

[2] O. R., xlv. part i. p. 1170.   [3] *Id.*, p. 1106.

hastened his movement by a single day, Thomas's concentration would have been balked by Hood's intercepting the railway trains that were bringing Steedman's two divisions from Chattanooga to Nashville.

The position on the north side of Duck River at Columbia was a difficult one to hold against a superior enemy determined to force a crossing. The river makes a large bend to the southward, enclosing a tongue of land considerably lower than the banks on the other side. It was just such a place as the enemy approaching from the south would choose for forcing the passage of the stream. Artillery could be placed on either side of the bend so as to cross-fire upon the bottom land and enfilade any intrenchments or lines of troops facing the crossing place at the apex of the bend. The bridges had been destroyed and the river was high, but falling. Behind the low land in the bottom, however, a ridge rises to a height equal to the south bank of the river, and this made a strong position in which to resist a hostile advance northward, even if the crossing were accomplished. Schofield did not feel much concern, therefore, as to his ability to hold these heights; but the river was falling and fords might already be available, above or below, by which Hood could turn the position. Then the advance, when we should be prepared to resume the aggressive, could not be made at this point, because the features of the position gave the enemy such a command of the crossing as to make it impregnable against a movement from the north. It seemed wiser therefore to retire a little farther and find ground from which we should have a choice of roads by which we might march against the enemy when our reinforcements should come.

In the forenoon of the 28th, Schofield sent dispatches to General Thomas,[1] discussing the situation and suggesting the concentration of his own forces so that he should be ready for prompt movement. This had special reference to the withdrawal of Brigadier General Cooper's troops from Centerville, thirty miles down the river, where they had been stationed by Thomas in the belief that Hood was most likely to advance upon Nashville by that route.[2] Early in the afternoon news came to Schofield[3] that the enemy's cavalry under Forrest had crossed the river near the Lewisburg Turnpike, some twelve miles eastward, and had pushed back our horsemen, under General Wilson, beyond Rally Hill. They had thus put themselves on our left flank, but it was not certain at nightfall whether Hood was using his infantry in this turning movement.

In sending this information to Thomas,[4] Schofield (at 4 P. M.) put the direct question where he proposed to concentrate and fight if it should prove true that Hood was moving in force upon our rear by the route his cavalry had taken. He also suggested sending a pontoon bridge from Nashville to Franklin to make

---

[1] O. R., xlv. part i. p. 1106.

[2] General Joseph A. Cooper (of East Tennessee) had temporarily commanded a division in the Twenty-third Corps. On our reaching Nashville, coming from Georgia in the early days of October, he had been sent with parts of two brigades and a battery of artillery to Johnsonville on the Tennessee River to assist in repelling Forrest's raid in that valley. After our concentration at Pulaski had begun, Cooper had marched with these troops to Centerville and Beard's Ferry (see map), by orders from General Thomas direct. O. R., xlv. part i. p. 1007. Meanwhile General Ruger had been transferred to the corps, and at Columbia a division was provisionally organized for him, consisting of Colonel Moore's brigade and a temporary one under Colonel Strickland, the latter made up by adding to Strickland's 50th Ohio other regiments which came up as reinforcements. See chap. iii., *post.*

[3] *Id.*, p. 1109, 1110.   [4] *Id.*, p. 1107.

good a crossing of the Harpeth at that place, the wagon bridge having been burned in one of the skirmishes of the campaign, and the river being too high to be easily forded.

The dispatches written by Thomas during the evening all attest his continued reluctance to consider the necessity of further retreat, and his strong wish that Schofield should hold the Duck River line till Smith's arrival; but none of them reached Schofield until eight o'clock next morning, and one written at half past three in the morning of the 29th did not reach its destination at all, but was captured by the enemy on its way.[1] This was the one in which definite instructions were given to retire to the Harpeth, in answer to Schofield's question of 4 P. M. of the 28th.[2] Schofield was thus left without orders, but

[1] Hood's Report, O. R., xlv. part i. p. 653.

[2] *Id.*, p. 1137. The importance of this dispatch, and the value of its information to the enemy, warrant me in giving a full copy of it. The Schofield dispatch of 4 P. M. seems to have been telegraphed from Franklin at 9 P. M., and is referred to by General Thomas as of that hour. See note 2 on next page.

"NASHVILLE, November 29, 1864, 3.30 A. M.

"MAJOR GENERAL SCHOFIELD, near Columbia, —

"Your dispatches of 6 P. M. and 9 P. M., yesterday, are received. I have directed General Hammond to halt his command at Spring Hill, and report to you for orders, if he cannot communicate with General Wilson, and also instructing him to keep you well advised of the enemy's movements. I desire you to fall back from Columbia and take up your position at Franklin, leaving a sufficient force at Spring Hill to contest the enemy's progress until you are securely posted at Franklin. The troops at the fords below Williamsport, etc., will be withdrawn, and take up a position behind Franklin. General A. J. Smith's command has not yet reached Nashville; as soon as he arrives I will make immediate dispositions of his troops, and notify you of the same. Please send me a report as to how matters stand upon your receipt of this.

"GEORGE H. THOMAS,
*Major General U. S. Vol's.,
Commanding.*"

under the influence of a long series of dispatches, every one of which breathed the strong desire that he should stubbornly resist the enemy's further advance. He resolved to await positive knowledge of the movement of Hood's infantry, and to risk making successful manœuvres in retreat when his opponent should be on the north side of the river. The result proved his capacity as well as his courage.

The sun rose about seven o'clock on the morning of the 29th, and a little earlier than that hour Schofield had received by courier a dispatch from General Wilson that was a duplicate of one written at one o'clock, which seems to have miscarried. The duplicate had a postscript written at three in the morning, and was sent from Hurt's Cross Roads on the Lewisburg Turnpike, by way of Spring Hill, a route of more than twenty miles. This dispatch told of information procured from prisoners,[1] showing that the whole of Forrest's command was in Wilson's front, on the north side of the river, and that pontoons were being laid for the passage of infantry.[2] Information from prisoners is apt to be unreliable, and is sometimes purposely deceptive; but Wilson seemed so confident of the truth of this that Schofield determined to act upon it, by putting his trains and three divisions of infantry at Spring Hill, and keeping Wood's and my own in position during the day, if possible, or until he should get answers from Thomas

[1] O. R., xlv. part i. p. 1143.
[2] General Schofield has put on file in the War Department complete copies of all dispatches which passed between General Thomas and himself in November, 1864, annotated with notes giving times of sending and of receiving, as determined by all available public and private memoranda, as well as by a careful comparison of the contents of the dispatches themselves. Nearly every question as to their order, the delays or failure in transmission, etc., is thus conclusively solved.

to his dispatches of the previous evening. Whilst these orders were preparing, the enemy opened with so lively an artillery fire from the town, and from the high banks encircling the tongue of bottom land, as to prove that most of his artillery was still in our front, with part, at least, of his infantry in support. Messages to Wilson were too slow, and it was evident that Forrest could prevent our cavalry from getting more definite news, and a strong infantry reconnoissance was determined upon as a preliminary before deciding finally to leave the line of Duck River.

About sunrise, Schofield therefore ordered Stanley to march with all of his command, except Wood's division and two batteries of artillery, to Spring Hill.[1] The trains were ordered to the same place, and Ruger was also directed to call in his smaller detachments down the river, and to move the parts of his division present, except one regiment at the ruins of the railway crossing (Ducktown), which was to remain as an outpost till night. My own division held the tongue of land in the bend of the river, covering the crossing by the turnpike, near which was a ford barely passable at a little lower stage of water. We were ordered to hold on till night, and then, leaving pickets till midnight, to follow the movement to Spring Hill. Wood was directed to send a brigade on a brisk reconnoissance in force up the river till it should develop the truth in regard to the actual crossing of the enemy's infantry. Captain Twining, Chief Engineer on General Schofield's staff, accompanied the reconnois-

[1] Spring Hill, eleven miles north of Columbia on the Franklin Turnpike, was at an important cross-roads, directly connected also by a diagonal road with Rally Hill, where our cavalry concentrated. It was therefore the first rallying place for the army in a retrograde movement.

sance, and reported its progress.[1] These orders were given, partly by notes sent through staff officers, and in part by conferences of General Schofield with his principal subordinates in person.

He had hardly given them when he received General Thomas's reply to his dispatches sent about nine o'clock of the previous forenoon, before he got the news from Wilson that Forrest was over Duck River.[2] This reply expressed the wish that he should retain his position at Duck River[3] till the arrival of Smith with his reinforcements, if he were confident of his ability to do so. It gave permission to have the distant detachments of infantry relieved, but directed that they should obstruct the crossings of the river by felling timber, and do this thoroughly as far down stream, on our right flank, as Centerville. The same courier brought an answer to Schofield's dispatch of 3.30 P. M., in which he had announced the first news of Forrest's crossing the river.[4] He had added the rumor that infantry was with Forrest, but said that he did not regard it as very probable. The report was, in fact, premature. Thomas's answer is couched in language which so vividly shows his frame of mind, and the conditions on which alone he could bring himself to think of further retreat, that it must be given in his own words: —

"Your dispatch of 3.30 just received. If Wilson cannot succeed in driving back the enemy, should it prove true that he has crossed the river, you will necessarily have to make preparations to take up a new position at Franklin behind Harpeth, immediately, if it becomes necessary to fall back."[5]

[1] O. R., xlv. part i. pp. 1139, 1140, 1141, 1142.
[2] Id., p. 1106. [4] Id., p. 1107.
[3] Id., p. 1108. [5] Id., p. 1108.

On the receipt of these dispatches, Schofield resolved to take the risks of a more deliberate movement. He countermanded Ruger's order to march,[1] and directed him to make thorough obstruction of the fords and roads down the river, in accordance with Thomas's instructions. A messenger was sent to Cooper at Beard's Ferry, near Centerville, ordering him to march to Franklin.[2] Kimball's division was ordered to halt at Rutherford's Creek, three or four miles in rear of Wood's position. The trains, however, were pushed onward to Spring Hill, where Stanley went in person with Wagner's division. Schofield informed Thomas that he would await the result of the infantry reconnoissance, and asked him to send orders directly to Cooper, as it was doubtful whether his messenger could reach Centerville.[3] This closed communication between Thomas and Schofield till the latter reached Franklin next morning; but when the former knew all the circumstances, he expressed the warmest approval of the conduct of his subordinate.

During the week preceding the 29th, the greatest source of embarrassment had been the organization of the telegraph service. The Secretary of War had, a year before, made the telegraph so far independent of commanders in the field that not even General Grant was allowed to have the key of the cipher used in his dispatches. This naturally led to great uncertainty in the service, the operators becoming lax in the performance of the task of ciphering and deciphering the dispatches which came into their hands. In the crisis of this campaign the telegraph was practically of no use, for a special messenger could have carried dispatches more speedily than they were transmitted by the wire. As early as the 20th of November, Scho-

[1] O. R., xlv. part i. p. 1142.   [2] *Id.*, p. 371   [3] *Id.*, p. 1137.

field had been obliged to ask why it should take all day to communicate between Pulaski and Nashville.[1] Again, on the 27th he informs General Thomas that he had just got a translation of the cipher dispatch of the 25th, two days after it was written.[2] A comparison of the times of writing and receipt will show that six or eight hours was the common interval. To cap the climax of embarrassments, the operator at Schofield's headquarters became alarmed for his personal safety on the 27th, and deserted his post, going to Franklin and refusing to return when ordered. He seemed to think it was a sufficient performance of his duty to decipher dispatches there and send the translation by a courier twenty miles. The security of the cipher was thus used over the line from Nashville to Franklin, which was perfectly safe, and the insecurity of sending the translation by a courier was suffered upon the route through Spring Hill in the presence of the enemy, and where, in fact, the most important dispatch of the series was captured by them.[3]

But this same dispatch had begun its misadventures at General Thomas's headquarters. It was written at half past three in the morning, but had not been sent to Franklin at six, and the recorded explanation is proof of the looseness of administration.[4] Why the Secretary of War had insisted upon such a system, in spite of the protests of military commanders, is one of the mysteries of the history of the war. In this instance it operated for the benefit of the Confederate general alone. The only rational plan was to put the key of the cipher in the hands of a confidential aid of

[1] O. R., xlv. part i. p. 958.
[2] *Id.*, p. 1086.
[3] See the facts as brought out in court martial proceedings, O. R., xlv. part ii. p. 289, and *Id.*, part i. p. 1172.
[4] O. R., xlv. part i. p. 1137.

the general to whom the dispatches were addressed, and limit the operator of the telegraph to the mechanical work of transmission. If a courier had to be used where the wire could not reach the general's tent, the dispatch would at least be unintelligible to the enemy that might capture it.

But let us return to the progress of events in Schofield's little army. At about eleven o'clock Captain Twining had reported a column of the enemy's infantry moving northward, on what proved to be the road from Davis's Ford (five miles above Columbia) to Spring Hill.[1] The discovery was mutual, for our column was seen by the enemy, and raised doubts in Hood's mind which grew more serious as the day wore on. The Confederate commander had marched with the greater part of his army, leaving General S. D. Lee in Columbia with his corps, diminished by Johnson's division, with orders to detain Schofield by an active demonstration until the afternoon. He was then to push seriously his effort to cross whilst Hood himself should attack our left flank.[2] The trains, and all the artillery except two batteries were with Lee, so that Hood might not be encumbered in his march upon the by-roads.

About noon Forrest's cavalry reached Spring Hill, having driven Wilson back toward Franklin on the Lewisburg Turnpike, and separated him from connection with our infantry.[3] Stanley reached Spring Hill with Wagner's division and most of the Fourth Corps artillery at about the same time with Forrest, and easily checked his advance. The trains which were just arriving there under the escort of a regiment from my division (103d Ohio) were parked; the artil-

---

[1] O. R., xlv. part i. p. 1139.
[2] Lee's Report, *Id.*, p. 687.   [3] *Id.*, p. 763.

lery was placed on commanding ground and positions were taken to cover the approaches to Spring Hill from the south and east. Stanley was also ordered to communicate with Wilson, and tell the latter not to let Forrest get between him and the infantry.[1] It was too late, however, to accomplish the last direction, for the whole of our cavalry was now on the Lewisburg and Franklin road several miles to the eastward, and our communication with it was not renewed till we reached Franklin next morning.

About three o'clock Schofield became satisfied that Hood was not intending to make his attack on Kimball's and Wood's divisions, which were in echelon on my left and rear. He therefore led Ruger's two brigades in person by a rapid march to Spring Hill, leaving a regiment of that command to guard the river at the broken railway bridge, as has already been mentioned.[2] As he passed Rutherford's Creek he ordered a brigade of Kimball's division also to follow him, for by that time the noise of the combat at Spring Hill told of a vigorous attack by the enemy. In anticipation of the probable course of events, written orders had been prepared at Schofield's headquarters directing a continuous movement toward Franklin, beginning at dark, and these had been distributed before Schofield went himself to Spring Hill. They were necessarily modified in several particulars to meet the changing events of the afternoon and evening.[3]

The trains did not leave Spring Hill for Franklin until my division passed at midnight and took the ad-

[1] O. R., xlv. part i. p. 1141.
[2] *Ante*, p. 27.
[3] O. R., xlv. pp. 1139, 1140. The order in General Schofield's order book is dated at Spring Hill, but its issue was from his headquarters at Duck River early in the afternoon. He issued no written orders at Spring Hill.

vance. Ruger's division was moved under Schofield's oral orders, and was the first to reach Stanley. The orders for the cavalry could not take effect, for the two brigades which Schofield supposed were at Spring Hill had joined Wilson, and when he arrived at that place were out of reach of orders. The organization of the column had to be changed in all these respects, therefore, and it was ordered by General Schofield orally and in person.

It is not my purpose to give the details of Stanley's sharp engagement at Spring Hill, or of my own bickering combat with Lee's corps at the river. I have already given these in another place.[1] My present aim is to show what new light we may get from the publication of the official records upon General Schofield's reasons for his action in the conduct of the retreat upon Franklin. These records give us the means of accurately knowing the situation from hour to hour, and the dispatches received by him, as well as those which miscarried.

Hood's attack upon Spring Hill failed, and Lee did not succeed in crossing the river at Columbia. At nightfall my division was withdrawn from its position at the river,[2] leaving the pickets in position supported by the 12th and 16th Kentucky Regiments, the whole under command of Lieut. Colonel Laurence H. Rousseau of the former regiment, accompanied by Major T. T. Dow, Inspector General on my staff.[3]

The picket lines were maintained till past midnight, and were then withdrawn without molestation. The

[1] See March to the Sea, Franklin and Nashville, chap. iv.
[2] O. R., xlv. part i. p. 1143.
[3] In my history above referred to, I said that Lieut. Colonel J. S. White of the 16th Kentucky was in command. Both these gallant officers were present, but I have learned that Colonel Rousseau was the senior.

enemy was not aware of this till about half past two, when Lee's corps began its preparation for following us.[1]

When Schofield reached Spring Hill with Ruger's division, darkness had put an end to the fighting there; but he learned that some force of the enemy was upon the turnpike at Thompson's Station, three miles north. Cheatham's corps lay in front of Stanley, his camp-fires within half a mile of our road, seeming much nearer in the darkness. Indeed, Schofield's own escort, moving in advance of Ruger, had brushed away a hostile picket from the road itself as he drew near to Spring Hill.[2] The brigade from Kimball's division accompanying him (Whitaker's) was ordered to extend Stanley's line to the right, giving better cover to our line of march. When Schofield learned of the presence of the enemy on his line of communications, and that our cavalry was wholly beyond reach, he marched with Ruger at once for Thompson's, leaving with Stanley directions for me to halt at Spring Hill till I should hear further from him.[3] He found the road clear, however, the detachment of Forrest's men who had been there having retreated, leaving their camp-fires burning.

Schofield now posted Ruger's division there, and rode back to Spring Hill, arriving a few minutes after I myself had reached Stanley's quarters. It was now midnight. The troops were marching left in front, so as to face the enemy on our right with least delay, and as the movement was in retreat, I was with the rear of my column. The division had massed by the

---

[1] Lee's Report, *Id.*, p. 687.
[2] Letter of Colonel Wherry of Schofield's staff. The escort was Captain Ashbury's company, 7th O. V. Cavalry.
[3] Schofield's Report, O. R., xlv. part i. p. 342.

roadside as it came up, and moved again as soon as I had my interview with the commandant of the army. Wood's division was already following my march, and Kimball's was preparing to follow Wood. The pickets at Duck River were gathering upon Rousseau's support, and the infantry and artillery of our little army were steadily concentrating upon Spring Hill. Schofield ordered me to take the advance at once, and march to Franklin, still twelve miles distant.[1] The trains followed, convoyed by Ruger's and Wood's divisions marching by the side of the wagons; Kimball's division followed, and Wagner's acted as rear guard.

The pickets with their supports, under Lieut. Colonel Rousseau, reached Spring Hill at about four o'clock in the morning. Wagner's division was still there, and Rousseau, pressing vigorously forward, overtook General Wood's column near Thompson's Station, as day broke, and at his request joined his command in defending the trains against some spirited attacks of the enemy's cavalry, and marched the rest of the way in company with that division.[2] Wagner did not get wholly away from Spring Hill, till six o'clock but was not vigorously pursued.[3] His greatest embarrassment was with the stragglers. Some two thousand new recruits had joined the army during the halt at Duck River, and were not yet inured to the fatigue of marching. It was hard work keeping them up with the column, so that they should not be captured. Their

---

[1] The distance from Columbia to Franklin by the turnpike is twenty-three miles. My position at Duck River being a mile from Columbia, my whole night march was twenty-two miles. See Report, Appendix B.

[2] Colonel Dow's statement, Appendix E.

[3] Opdycke's Report, O. R., xlv. part i. p. 239.

knapsacks were carried for them or thrown away. Ambulances and wagons took those who were most footsore, and, though the rear of the column was delayed, they were safely brought into camp at Franklin before noon.

CHAPTER III

TAKING POSITION AT FRANKLIN

Arrival at Franklin — No Bridge or Pontoons — Hood to be held back — Schofield's Oral Directions — His Correspondence with Thomas — Delay in Arrival of Reinforcements at Nashville — Can you hold Hood back three Days? — Orders to continue Retreat — The Position at Franklin — The Carter House — The Town and the River — The Field in Front — The Defensive Line — Repairing Bridges — Twenty-third Corps Positions — Reilly's Division — Ruger's Division — The Works on Carter Hill — Retrenchment across Turnpike — Kimball's Division.

MY own march from Spring Hill to Franklin had been undisturbed. We had taken an easy gait, so as not to outmarch the trains and their guards, and it was about half past four o'clock, or two hours and a half before day, when we approached the outskirts of the village. General Schofield had ridden forward and overtaken me after he had issued his final orders at Spring Hill, for it was of the greatest importance that he should study the position north of the Harpeth, and the means of crossing that river. It was hardly less so that he should get into direct telegraphic communication with General Thomas. On reaching Thompson's Station in the night, he had sent Captain Twining, his engineer, with a small escort, to Franklin, to communicate with Thomas, and to examine the means of crossing the river. Twining was also instructed to order forward any troops of A. J. Smith's command that might have reached the Harpeth. He arrived at Franklin at one o'clock in the morning, and telegraphed Schofield's situation to General Thomas, and his expectation that he might not get beyond Thompson's Station that night. He added that Schofield thought it likely that he might be forced into a general battle on the 30th, or lose his wagon train.[1]

When the outermost house of the village came dimly into view as we marched northward, Schofield directed me to mass my division on both sides of the turnpike, leaving the way clear for the trains, and let the men make their coffee, whilst he rode into town to find Captain Twining, and learn the condition of the river crossings. He was especially anxious to know whether the pontoons had arrived which he had asked for by a second and urgent dispatch.[2] To Schofield's first request (sent on the 28th), Thomas had answered by giving authority to use some of those which had been at Columbia;[3] but the fact was that those had been very heavy wooden boats, for which there were no trucks large enough to carry them up to the railway, or proper cars for their transportation, and they had been destroyed when the crossing was abandoned. The second despatch (dated at 1 P.M. on the 29th) reached Spring Hill after the enemy had possession of the road northward, and could go no farther. Not being informed of this, Schofield was hopeful that he would find a bridge train awaiting him at the river.

[1] O. R., xlv. part i. p. 1138. Twining's dispatch as printed in the Records is dated at 10 o'clock P. M. of the 29th, but General Thomas in replying (p. 1168) speaks of it as "your dispatch of 1 A. M. to-day," i. e. the 30th, and this best accords with other events. Schofield was at Thompson's Station at about eleven, and returned in haste to Spring Hill, where he met me, whilst Twining was on his way to Franklin.

[2] *Id.*, p. 1138.   [3] *Id.*, pp. 1107, 1108.

After giving orders for resting the troops, I rode on with my staff to the house before me which was on our left hand as we approached the town, and was partly hidden by a grove of trees a little way south of it. Rousing the family, they were told that we should have to make use of part of the house as temporary headquarters. They put their front sitting-room at our disposal, and, loosening sword belts and pistol holsters, we threw ourselves upon the floor to get a few minutes of greatly needed sleep. I had fallen into a doze when General Schofield returned.

In all my intimate acquaintance with him, I never saw him so manifestly disturbed by the situation as he was in the glimmering dawn of that morning. Pale and jaded from the long strain of the forty-eight hours just past, he spoke with a deep earnestness of feeling he rarely showed. "General," he said, "the pontoons are not here, the county bridge is gone, and the ford is hardly passable. You must take command of the Twenty-third Corps, and put it in position here to hold Hood back at all hazards till we can get our trains over, and fight with the river in front of us. With Twining's help, I shall see what can be done to improve the means of crossing, for everything depends upon it. Let your artillery and trains go over at once. I will give you batteries from the Fourth Corps, in place of yours, as they come in." So vivid is this recollection, that, as I recall it, I seem to hear the very words. The written orders which were issued when Major Campbell, the Adjutant General, established his office later in the day, were only the formal embodiment of the purpose thus expressed. On this I acted. It will become necessary, by and by, to examine the text of the orders, and note the points of discussion which have arisen concerning them. This is the reason why I state so fully the manner in which General Schofield's purposes and commands were made known to me by himself.[1]

On his first arrival at Franklin, General Schofield found Captain Twining, and read General Thomas's dispatch of four o'clock, giving the news that A. J. Smith's command was upon boats at the levee.[2] Thomas thought, however, that it would be impracticable for Smith to reach Franklin that day, and indicated a wish that Schofield should get into position on the north side of the Harpeth, and continue the efforts to delay Hood. Schofield's answer is dated at five o'clock, and urges that Smith should march at once, as he could, at least, protect the wagon trains if further retreat became necessary.[3] A half-hour later he reported his hope to get his troops and trains over the river during the morning,[4] and that he would try to get Wilson, with the cavalry, upon his flank again. A little before ten, he informed Thomas that half his troops were in, and the other half about five miles out, coming on in good order; that Wilson had joined him, and was in position on his flank, but could not cope with Forrest, by whose help Hood could cross the river

---

[1] Major C. S. Frink, Brigade Surgeon and Medical Inspector on my staff, in a letter of June 26, 1881, relating to the hospital service during the battle, incidentally said that he judged a battle to be imminent "from the conversation I overheard between you and General Schofield when he formally placed the command of the whole line in your hands." He later reaffirmed this, referring for corroboration of his memory to a letter he had written home the day after the battle, and which was before him. At the time of this correspondence, in 1881, I was collecting material for my historical volume already mentioned.

[2] O. R., xlv. part i. p. 1168.
[3] Ibid.
[4] Id., p. 1169.

whenever he should seriously attempt it. He therefore asked whether he should hold on till compelled to fall back.[1]

We must not forget that Schofield had heard nothing of his cavalry since the middle of the forenoon of the 29th. He was relieved of some anxiety, therefore, when he found at Franklin a dispatch from the cavalry commander, dated at ten o'clock in the evening, from a point two and a half miles from Franklin, on the road to Triune (east).[2] Wilson had ordered Hammond's brigade to march to the latter place (twelve miles) without halting, and report all movements of the enemy in that direction. One of Schofield's first acts after sending his earliest dispatch to Thomas was to send directions to General Wilson[3] to cover his immediate flank and rear during the day, with at least a portion of his cavalry. He felt that the absence of his mounted troops had added greatly to the perils of his situation during the retreat from Columbia.

It will tend to clearness to dispose at once of the subsequent correspondence between Thomas and Schofield on this day. A dispatch from the former, at 10.25 A.M. (written before the receipt of Schofield's of 9.50), stated that it would take Smith's troops the whole day to disembark, but he would then send them to Franklin or to Brentwood (half way), unless he found it necessary to keep them at Nashville.[4] This proviso refers to his appehension

[1] O. R., xlv. part i. p. 1169. In General Schofield's report this dispatch is correctly dated at 9.50 A.M. General Thomas's report has it 12.30 M. By the latter hour, Wagner's division was in position on Winstead Hill, and the rest of the army and the trains had passed the lines in front of Franklin.
[2] O. R., xlv. part i. p. 1145.
[3] Id., p. 1177.
[4] Id., p. 1169.

that Forrest would make a raid upon Nashville. Wilson's dispatches to Thomas had been full of this idea, which, though a mistaken one, was manifestly influencing Thomas's action. Schofield was therefore told that Franklin should be held till the problem might be cleared up a little, unless this should involve "too much risk." Of course the whole gist of the question was, What is too much risk? Schofield answered at noon [1] that he thought he had already been running too much risk, using the language already quoted; that the slightest mistake on his own part or failure of a subordinate during the last three days would have proved disastrous.[2] Yet he promised the most cheerful obedience in carrying out his superior's views when they should be distinctly expressed. He urged concentration, if not at the front, then at some point further in the rear. His question whether he was to hold fast "until compelled to fall back" had not been answered. A little later Thomas learned that one division of Smith's forces had not arrived. He telegraphed Schofield that they would have to try to hold Hood back till these troops could come.[3] "After that," he said, "we will concentrate here, reorganize our cavalry, and try Hood again." He therefore asked Schofield whether he thought he could keep Hood at Franklin for three days longer. As Schofield had already expressed the opinion that Hood could cross the Harpeth whenever he attempted it, the question seems unnecessary; but Schofield replied, at 3 P.M., that he did not think he could insure three days.[4] One day he might answer for, because the manœuvres of the enemy would require about so much time when

[1] O. R., xlv. part i. p. 1169.
[2] Ante, p. 22.
[3] Id., p. 1170.
[4] Ibid.

CARTER HILL

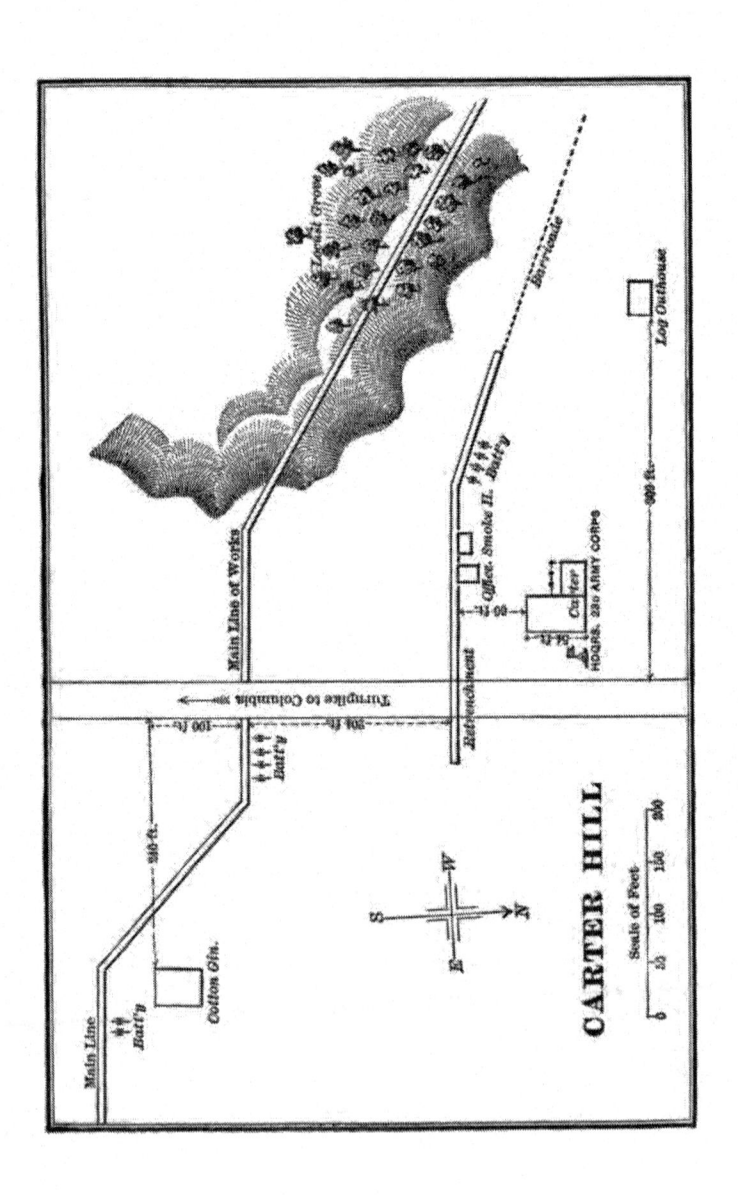

a new flanking movement should begin. In fact, preparations to cross above and below were at that time making, and skirmishing between Forrest's and Wilson's forces, three miles above, had just been reported. Schofield's opinion, which Thomas had asked for, was that he should, in view of Thomas's decision that Smith's forces could not leave Nashville that day, take position at Brentwood, and that Smith's division, which had arrived in the night, together with the Murfreesboro garrison, should join him there.[1] In this, so far as Schofield's own movement was concerned, Thomas acquiesced, and directed the wagon trains to be put on the road at once. Hood's attack came to cut the knot, probably before Schofield received the last despatch, and the battle materially changed the situation. This correspondence exhibits the frame of mind in which the two generals were, and emphasizes the purpose which Schofield expressed to me at the Carter house before the break of day, that my duty was to use the forces put under my command to hold Hood back, at all hazards, until the trains and the rest of the army should be safely across the Harpeth.

The house at which I had stopped belonged to Mr. F. B. Carter, an aged man, who occupied it with several grown children and near relatives. A son was a Confederate officer, who had been taken prisoner, and was now at home on parol.[2] The house

[1] O. R., xlv. part i. p. 1170.
[2] The younger man, Mr. M. B. Carter, is now the owner of the farm, and still resides there, and has been most courteous and helpful in fixing the important points of location and topography connected with the battle-field. He himself levelled the breastworks, and as he has continuously lived upon the spot he is a conclusive authority on some interesting questions in regard to which accounts have differed. See the account of a visit to the field in 1888 by Captain Thomas Speed,

was of brick, of ample dimensions on the ground, but not very high, as it was built of one lofty and airy story, with attics above. An ell extended the suite of rooms on the north side of the hall some thirty feet toward the rear. On the south side of the door-yard, sixty feet south of the house, and opposite the ell, two smaller buildings stood. The one nearest the road was of wood, and was used as an office; the other was a brick smoke-house, and the passage between the two was ten or a dozen feet wide. The house itself stood forty-five feet back from the road, which was bordered by a line of shade trees. A barn with a log corn-crib and some smaller outbuildings were a little over a hundred yards back from the road, and some thirty yards farther north than the line of the house. This group of buildings was to become the focus of so desperate a conflict that it is well to have their situation and relative position clearly understood.[1]

As soon as it was light enough to select positions for the troops, I carefully examined the ground. Looking northward toward the town, a well marked slope leads to a lower level on which the place is built, the public square in its centre being forty feet lower than the knoll or bench at the Carter house. The pretty village itself is a third of a

"Sketches of War History," vol. iii. p. 44; published by the Ohio Commandery of the Loyal Legion. In company with Captain Speed and Major D. W. Sanders (formerly of the staff of General French, who commanded a division in Hood's army), I visited the field recently and made careful verification of the topography.

[1] For the accurate measurements about the Carter house I am indebted to Colonel M. B. Carter, who is a practical surveyor. The office building has, since the war, been moved from the place it occupied in the battle, and is now an extension of the ell of the dwelling-house. In moving it, it was turned about so that its south side, which is riddled with bullets, now faces northward, toward the town.

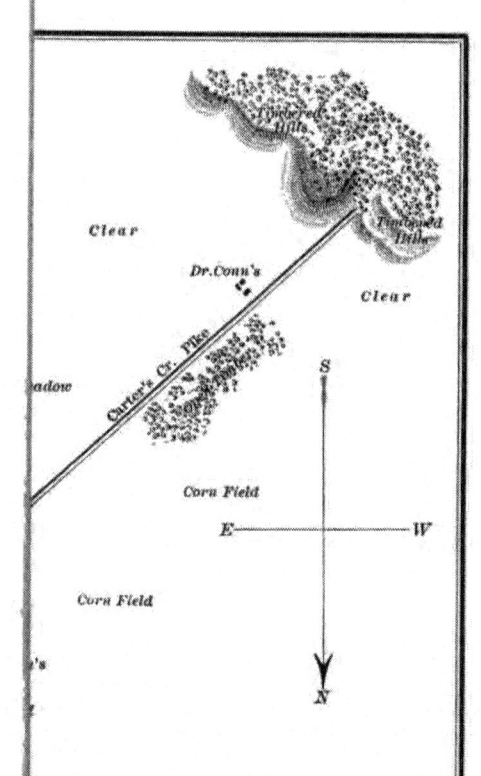

FIELD WORKS
AT
FRANKLIN, TENN.

Occupied by the 23rd and 4th Corps
during engagement of Nov. 30th, 1864
Maj. Gen. J.M. SCHOFIELD, Com'd'g.

Official
Wm. J. Twining,
Ch'f Eng'r, Army of the Ohio.

Scale of Yards
0 50 100   200   300   400   500

mile away, an open belt of fields and gardens then encircling it from river to river as it stands in the deep re-entrant angle of the Harpeth. The bend of the river is almost a right angle, and the stream washes the east and north sides of the town. As one goes up stream, he finds, after getting beyond the houses, that the valley turns to the southeast; but down stream it trends to the northwest, which is the general course of the river till it falls into the Cumberland, some thirty miles away. In the re-entrant angle the northerly bank is the commanding one, being not only of a higher general level, but having also well marked hills, on one of which (Figuer's Hill), enfilading the stream and railway on the eastern side of the village, was Fort Granger, a dismantled earthwork, built a year or two before. The railway bridge is perfectly covered by any artillery placed in the fort, and a deep cut in the railway at the edge of the town may also be thoroughly swept from that position.

The streets in the town are not in the same direction as the turnpike at the Carter house. When the highway from Columbia enters the village, it turns to the right to reach a favorable place at the river for the bridge, and the squares are symmetrical with this line.[1] The turnpike bridge had been destroyed early in the war, and had not been rebuilt. It was a single span of covered wooden truss, resting on high abutments. What was in 1864 known as the County Bridge was a lower and cheaper structure on trestles, built near the railway crossing, where a hollow on the north side made a practicable ascent for the roadway. This departure from the straight line of the turnpike added nearly half a mile to the

[1] See Map.

length of the road, besides making it more difficult by reason of the grades. The ford was between the site of the former turnpike bridge and the county bridge. From the Carter house through the village to the ford is about a mile.

Turning now to the south, from the same point of view, it is seen that the Columbia Turnpike is nearly level, rising slightly till it crosses a low summit half a mile distant, and then dipping again so as to hide men or teams in the road. Most of the space to the Winstead Hill, two miles away, is so gently undulating as to look like a plain with a few depressions in it on right and left of the central ridge and road, where small watercourses run either way to the Harpeth. About half way to Winstead Hill, a bold, stony hill rises on the west of the turnpike, isolated in the general level around it, and known in the neighborhood as Privet Knob.[1] Winstead Hill bounds the valley on the south, making part of a circle of ridges and heights, which seem to surround the basin in which the town lies. The Columbia Turnpike runs straight south, lost to view after it passes the low summit above mentioned, but coming into view again, climbing Winstead Hill by a white line rising from left to right, and passing over the crest between two of the rounded summits, which give the elevation a picturesque outline.

Two other turnpikes run fom the town southward. That to Lewisburg goes up the Harpeth valley in a southeasterly direction. The other is at the west, and leaves the town by a similar angle. It is called the Carter's Creek Turnpike. The map of Franklin

---

[1] In some of the reports of the battle this hill is called Stone Hill; in others, Merrill's Hill. Privet Knob is the local name given it in a former generation, from the privet thicket which covered it.

and its surroundings has been aptly compared to the left hand extended with separated fingers. The little finger and thumb at right angles represent the Harpeth River in its course from left to right, whilst the three fingers spread in the midst indicate the three turnpikes diverging southward from the village.

Half a mile southwest of the Carter house, and near the Carter's Creek Turnpike, is a hillock with a mansion and orchard known as the Bostick place.[1] Between the two houses is a gentle hollow, which is about thirty feet below the level in the direct line from house to house. In it heads a small watercourse, which meanders through it, and, crossing the Carter's Creek Turnpike, curves northwardly to the Harpeth. This hollow, with its marshy brook, bounds the village on the west.

Looking eastward, the Carter Hill went forward a little, and one then saw, a hundred and twenty yards in front and eighty yards east of the turnpike, a cotton-gin, a strong frame building like a barn on the most advanced salient of the hill. To the left of this the ground descended a little, but rose again on reaching the Lewisburg Turnpike, half a mile away, where, between it and the railroad and a little in the rear, was another well marked knoll, through which the railway excavation cuts, as has been already mentioned.[2] In the middle of this recurved line, on the left, were two large oak trees, still standing, landmarks which show the curve of the natural lines of defence. Beyond the knoll and the railroad was

[1] In the Twining Map (p. 45) it is marked as a group of buildings in a square enclosure. The name F. B. Carter, near it, is that of the owner of the large surrounding farm, but he did not own this mansion.
[2] *Ante,* p. 45.

the upper reach of the river, widening the field in front as it bore off to the eastward.[1]

Such was the field as it lay before us under the level beams of the rising sun. It was evident that the Carter Hill was the key to any strong system of defence in front of the town, and that the line from the cotton-gin as a salient by the oaks to the knoll near the river bank must form our line of battle on the left flank. Artillery in Fort Granger, on Figuer's Hill, would be a powerful support to it. As to the right flank, a gallop to the Carter's Creek Turnpike showed that it would not do to throw that wing out to the Bostick place, although that position was the most commanding one in itself. Our flank resting there would be "in the air," with no security against its being readily turned by the enemy. Having but two divisions of the corps in hand, we must find a shorter line, and let the hollow with its watercourse determine the outline of our breastworks in that direction. Our position would, in part, be lower than the Bostick Hill, and commanded by it, but we could assist our troops there by an artillery cross-fire from the Carter Hill. I hoped we might have reserve enough to make a second line on the extreme right. At any rate, it was the best we could do, and if we could hold back the enemy during the day, we should be at liberty to put the river between us and him at nightfall.

The situation, as I have described it, and the general topography of the region, made it probable that Hood, after passing Winstead Hill, would push his right flank forward on the shortest line to our

---

[1] The knoll on our extreme left is now occupied by the residence and grounds of Mr. Rolffs, a merchant of Franklin. There was no house on it at the time of the battle.

communications with the north bank of the Harpeth, for both the bridges, the wagon bridge and that of the railway, as well as the ford, were in rear of the left of our defensive line. On the upper reaches of the river were also the nearest and most available fords by which his cavalry, under Forrest, could cross and turn our position. These considerations decided me to put my own division (the only troops then in hand) on the line from the Carter house toward the left, making the knoll near the river strong with a recurved line, which should guard the railway cut. The most essential part of our defences would thus be first prepared, and Ruger's division, as it should come in, could extend the lines westward with a refused flank at the Carter's Creek Turnpike.

Let us now go back and follow the actual course of events from the time General Schofield gave me his general directions, and left me in command at the Carter house, whilst he went, in the morning twilight, to give personal supervision to the all-important work of improving the means of crossing the river. We had in the corps an efficient engineer battalion, made up of intelligent mechanics, and it was ordered at once to report to Captain Twining, Chief Engineer. The artillery of the corps under Captain Cockerill, Chief of Artillery, was ordered to cross at once by the ford, and as it reached the north bank Cockerill's own battery of three-inch rifles was placed in Fort Granger by General Schofield's personal order. The rest of the cannon were parked near by.[1]

The trains were also ordered to proceed at once to the town, and prepare to follow the artillery across

[1] O. R., xlv. part i. p. 432.

as soon as possible. The approaches to the ford had to be mended by scarping the bank on either side, so as to make a practicable grade for our heavy laden army wagons. Most of the Twenty-third Corps ammunition train also passed over to the north side, as the uniform calibre of the infantry arms enabled us to rely on the Fourth Corps trains, which came in last.[1]

The county bridge had been burned in a skirmish earlier in the season, but had not been wholly destroyed. It was found that, by sawing off the posts quite close to the water, and making new stringers and flooring beams for the intermediate spaces, a temporary structure of a useful character could be made. It was only intended at first for the passage of the troops themelves, but it was found to be so substantial that loaded wagons could be taken over it with a little care. Its floor was so near the water that many who crossed it thought it was a pontoon bridge, and it was so spoken of by several officers in reports and printed statements. They had heard that a pontoon bridge was expected, and naturally assumed that it had been laid.[2] The change of grade necessitated a good deal of work at the approaches of this crossing also, before it could be available for teams. But we had still a third

[1] Overlooking this fact, Van Horne speaks as if the great quantity of ammunition used from the Fourth Corps train were expended by the troops of that corps alone. Army of the Cumberland, ii. p. 202. It probably represents what was expended by all the troops engaged except the few companies having repeating rifles.

[2] After the publication of "The March to the Sea," etc., I learned from General Schofield that the pontoons he had telegraphed for arrived in the course of the forenoon by railway train, but the other bridges had by that time been put in such condition that he thought it best not to lay the pontoon bridge, and sent the train back to Nashville.

means of getting over the river,— the railway bridge. By dismantling some stables and sheds in the town this was planked, and by skilful grading at the ends was brought into use for wagons as well as for the marching troops.[1] It was late in the forenoon before any other crossing was practicable for wagons, except the ford. The wagon trains as they came in were parked in the cross streets, leaving the main thoroughfare open, and one by one they slowly, and with great difficulty, made their way through the ford. As soon as either of the bridges was available, more continuous lines of teams were set in motion; and when the battle opened, in the afternoon, most of our wagons were quietly wending their way toward Nashville. Nothing but the most intelligent and energetic use of the means at hand had made possible the saving of our trains by these improvised bridges, and nothing but strict discipline and system in the handling of the wagons by the quartermaster's staff and employees prevented confusion and consequent blocking of the way.

As soon as the troops of my own division had eaten their breakfast, Brigadier General James W. Reilly was put in temporary command of them, and was directed to intrench them in the position on the left of the Columbia Turnpike which I have already described.

The three brigades were marched at once upon this line, making their front conform to its angles, as our habitual custom was. Their arms were then stacked in rear, intrenching tools distributed, and each regiment ordered to cover its own front. Calculation was made to leave room for two batteries of artillery on the left of the turnpike, and two others

[1] Schofield's Report, O. R., xlv. part i. p. 342.

were promised me for position in the line on the centre and right.

Henderson's brigade (Colonel I. N. Stiles in temporary command on account of Colonel Henderson's illness) was placed on the extreme left, holding the line between the river and the Lewisburg Turnpike. Casement's brigade came next, occupying a straight line toward the cotton-gin, marked by the two oak trees which were left standing. Reilly's brigade was the right of the division, and had a short front from the Columbia Turnpike eastward, joining Casement beyond the cotton-gin. It will be remembered that two of Reilly's regiments (12th and 16th Kentucky) were the support of the picket line left at Duck River, and did not reach Franklin till noon. Reilly therefore made his front of two regiments only, — the 100th Ohio next the turnpike, and the 104th Ohio on its left, — the cotton-gin being in rear of this regiment. The 8th Tennessee (a small regiment of loyal men from East Tennessee) was placed in reserve at the centre of the brigade, leaving room in second line for Rousseau's and White's Kentuckians when they should arrive. A new regiment of Ohio recruits, the 175th Ohio, which had very recently arrived, had been assigned to duty in guarding the railway south of Franklin. Its detachments were collected as our columns withdrew, and when these assembled at Franklin, in the course of the morning, the regiment was temporarily assigned to Reilly, and placed in reserve. Casement's brigade was formed with the 65th Indiana on the right, with the 65th Illinois and the 124th Indiana completing his first line.[1] His reserve was the 5th Tennessee,[2] and his own regiment (103d Ohio), which was greatly reduced

[1] O. R., xlv. part i. p. 425. [2] Id., p. 429.

in number by casualties of the preceding campaign, was on detached duty as guard for the Twenty-third Corps headquarters and ordnance trains. Stiles's brigade completed the line, having the 128th Indiana next to Casement on its right, the 63d Indiana in the centre, and the 120th Indiana on the extreme left. He had also the 112th Illinois in reserve. Part of Stiles's front was just in rear of a hedge of Osage orange, which was thinned out so as to make an admirable palisade outside the ditch, and the material obtained from this, as well as from some other hedges near by, was used as an abattis on the rest of the line, nearly or quite to the Columbia Turnpike. It was too small in size, but it was tough and very thorny, and proved to be a useful obstruction, troublesome to meddle with under fire. We lacked timber for revetment of the earthworks, and consequently a ditch was made both inside and outside the parapet in many places. We succeeded in finding material to make the ordinary head-logs, and so before noon this division had a fair defensive line.

As soon as my headquarters wagon came in, two or three tents were pitched in the door-yard of the Carter house, on the slope toward the village. Captain Cox, my Adjutant General, opened his office there, and the headquarters flag was unfurled in its front. This was the centre of the line when Ruger's division came into its position, and it was during the whole day and night the point to which all communications came, both from the line itself and from General Schofield, when he sent orders or inquiries relating to this front, until the battle was over.

Ruger's division began to arrive about seven o'clock, and was assigned to position between the

Columbia and the Carter's Creek Turnpikes.[1] He had but two brigades present. Cooper's, having been at Centerville, was now making its way toward a crossing of the Harpeth River, several miles farther down the stream. Cooper had with him also two regiments of Strickland's brigade, and their place was temporarily supplied to Strickland by the 72d Illinois and the 44th Missouri, which belonged to the Army of the Tennessee, and had been brought from Memphis to Nashville in time to join Schofield's army at Columbia. The 44th Missouri was a new regiment, very recently organized from fresh recruits.[2] Another new regiment, the 183d Ohio, had also reached us a day or two before, and was, for the time, under Strickland's orders, whose brigade therefore contained only one regiment of the old troops of the corps. The two veteran regiments (50th Ohio and 72d Illinois) were put in front line on the right of the Columbia Turnpike, in the order named, and they were supported in second line by the 44th Missouri and the 183d Ohio. Moore's brigade was on the right of Strickland, and had to be stretched out in single line, without reserves, to enable it to reach and cover the Carter's Creek

---

[1] O. R., xlv. part i. p. 364.

[2] These two regiments have sometimes been spoken of as if they were the advance of General A. J. Smith's column. They had not been with him in Missouri, but were later assigned to his command. The 72d Illinois was part of the Seventeenth Corps, and had been left in garrison on the lower Mississippi, and brought to Nashville by General Thomas under his authority as Commandant of the Military Division of the Mississippi in Sherman's absence. It reached Columbia, November 22d (O. R., xlv. part i. p. 999), and was attached to Strickland's brigade (*Id.*, p. 1039). The other regiments which came from the Mississippi prior to the battle of Franklin may be traced in the same volume of the Records, pp. 1056, 1057, and 1084. The 44th and 24th Missouri were assigned to the Twenty-third Corps.

road.[1] Its order from left to right was as follows, viz.: 111th Ohio, 129th Indiana, 107th Illinois, 23d Michigan, 118th Ohio, and 80th Indiana. It was still too short, and two companies were detached from the 183d Ohio, and sent to Moore to be placed at his centre, between the 107th Illinois and the 23d Michigan.[2]

The breastworks built by Ruger's men were similar to Reilly's. The right and front of the southwest slope of the Carter Hill (looking toward the Bostick place) was covered with a grove of young locust trees,[3] and these, with an apple orchard near by, were used to make an abattis, though, like the material used for the same purpose on the other side of the turnpike, it was too light to be quite what was needed.

The position of the works from the cotton-gin to the barn and corn-crib in rear of the Carter house proved so important in the engagement, and became moreover the subject of so much controversy, that a fuller description of them should be given. As has already been said, the cotton-gin formed a marked salient in the line, the space between it and the parapet in front of it being only enough for the passage of troops and the working of the cannon which were in battery there. A little to the right of it the works made an angle toward the rear, coming back to join the épaulement for four guns on the left of the turnpike, ninety yards south of the Carter house. Where the line crossed the road, a gap

---

[1] O. R., xlv. part i. p. 351.

[2] The order of Moore's regiments is not given in the division or brigade report, and is made up from the reports of the regiments. The list in my own official report was incomplete. See Appendix B.

[3] See Twining Map, p. 45.

was left of the full width of the road, for the continuous lines of wagons and artillery crowded it all the morning. On the west of it, the line continued at right angles to the road for fifty yards on level ground, and then bent to the rear, descending the slope somewhat as it did so. This was with the purpose of placing a battery on the summit at the right of the brick smoke-house, which could fire over the heads of the infantry in the front line, and sweep the approaches in the direction of the Bostick place.

In rear of the opening in the front line, I ordered a retrenchment built across the road, and a turnout, so that the army trains could go around it on the left and regain the road. This retrenchment was in line with the south side of the office and smoke-house of the Carter place.[1] The retrenchment was continued to the right by Strickland's men, till it reached the small buildings. Space for a battery was left on the right of the smoke-house, and the men in the reserve line extended the infantry trench some distance farther. My personal knowledge of this was not distinct, as no second infantry line of trench had been ordered by me; for my purpose had been to keep the brigade reserves as a movable force to support the main line in case of need, and I did not wish the troops in the first line to be tempted to think they could leave it and fall back to another. But when soldiers at rest have intrenching tools they are apt to use them, and a sheltering ditch may be dug without any authority but that of a regimental or brigade commander.

My own recollection is, that about noon I had stopped at the centre of the 183d Ohio, the new

[1] See Sketch Map, p. 43. This is from a plat made by Colonel Carter, who verifies the position of the earthworks and retrenchment.

regiment on the right of Strickland's second line, to speak to Lieut. Colonel Clark, who had served in my command at the beginning of the war, when he was a member of the 7th Ohio, and who was to lay down his life heroically in this bloody struggle. As I recall the situation, the men of the 183d were then lying down, with no cover except the natural curve of the ground. The regiment next on its left was the 44th Missouri, and its commanding officer (after the fall of Colonel Bradshaw), Lieut. Colonel Barr, reported that it built breastworks on its line.[1] This was so fully corroborated by other official reports, and by my inspection of the field on our advance after the battle of Nashville, that I had no hesitation in stating, in my own official report of the battle, that here during the fight the troops, "under cover of the smoke, strengthened a barricade and breastwork that had been there before."[2]

Late in the forenoon the artillery of the Fourth Corps began to arrive, and, in accordance with General Schofield's promise, Captain Bridges, Chief of Artillery, reported to me with four batteries.[3] In his official report he says, "By direction of Brigadier General Cox, commanding Twenty-third Army Corps, at 12 M., I placed the 1st Kentucky Light Bat-

---

[1] O. R., xlv. part i. p. 395. [2] *Id.*, p. 354, and Appendix B.
[3] O. R., xlv. part i. p. 1172. The following is the text of the order: —

"HEADQUARTERS, ARMY OF THE OHIO, FRANKLIN, TENN.,
November 30, 1864.

"CAPTAIN BRIDGES, —

"The Commanding General directs that you report four batteries from your command to Brig. Gen. J. D. Cox for position on the line to-day.

"Very respectfully,
"J. A. CAMPBELL,
Major and Assistant Adjutant General."

58   *The Battle of Franklin*

tery, 6th Ohio Light Battery, 20th Ohio Light Battery, and Battery B, Pennsylvania Veteran Volunteers, in position in line, and had good embrasures made for their guns."[1]

I had marked places for these batteries, and they were placed thus: the Kentucky Battery, four guns, on the left of the Columbia road, in the line of the 100th Ohio Infantry, the 6th Ohio Battery, two guns in the salient at the cotton-gin and two guns on the left of the Lewisburg Turnpike, the 20th Ohio Battery on the right of the Columbia road just west of the Carter house, and the Pennsylvania Battery at the Carter's Creek Turnpike on Ruger's right.[2] A little later, when it became evident that the enemy was forming to attack in force, General Schofield sent Colonel Schofield, his Chief of Artillery, to meet Captain Bridges at my headquarters, with authority to use all the artillery that might be available.[3] Battery M, 4th U. S. Artillery, and Battery G, 1st Ohio, were placed on the extreme left, and Bridges's own Battery, Illinois Light Artillery, with Battery A, 1st Ohio, were held in reserve near the centre. The batteries were nominally six-gun batteries, but, in consequence of the hard work of the last two or three days, they were nearly all reduced to four guns each when reporting to me.

At noon our line of breastworks was completed on the left, and nearly so on the right of the Columbia Turnpike. The artillerymen were making embras-

[1] O. R., xlv. part i. p. 320.
[2] In the Official Records as published (vol. xlv. part i. p. 351) there is an omission by a clerical error in copying. The words "on the right of the Columbia," etc., are made to follow "6th Ohio Battery," omitting all that is between. My private copy is correct. See Appendix B.
[3] O. R., xlv. part i. pp. 321, 351.

ures for the guns of the four batteries already placed in the line, and were heightening and strengthening the infantry parapet to adapt it to this purpose. The guns of the Kentucky battery, close on the left of the turnpike at the centre, displaced three companies of the 100th Ohio Infantry, and Lieut. Colonel Hayes, their commander, placed these men immediately behind the battery, in close support.[1] Lieut. Colonel Rousseau also arrived at this time with the 12th and 16th Kentucky, the supports of the picket line at Duck River, and on reporting to General Reilly, these were placed in second line, fifty paces in rear of the first. Lieut. Colonel White, with the 16th Kentucky, had his right as close to the turnpike as the limbers and caissons of the battery would permit. The 8th Tennessee and 12th Kentucky extended this line to the left.

It will be seen that, in anticipation of a severe struggle on the line of the Columbia road, our lines were doubled there, and strengthened by artillery. On the extreme left, not only was the knoll in Stiles's brigade a strong position, but the artillery in Fort Granger, on the north bank of the Harpeth, flanked the position, and Cockerill's three-inch rifles had range enough to make their shell fire as effective as if they were in the line.

The only weak place was our extreme right; for though it was improbable that Hood's attack would extend farther in that direction than the Carter's Creek road, his cavalry might find the end of the line, and, unless there were some reserve there, might give us trouble. Indeed, our skirmishers, who were, as usual, well out on our front and flanks, reported at this time that a body of the enemy's cavalry was

[1] O. R., xlv. part i. p. 419.

reconnoitring on this road, and I called General Schofield's attention to the fact that Ruger's right flank had no secure point to rest upon.[1] Kimball's division of the Fourth Corps was just coming in then, and Schofield ordered it to report to me to strengthen this right flank.[2] I did not think it necessary or desirable to extend our proper front beyond the Carter's Creek Turnpike, and directed General Kimball to put his brigades in echelon on Ruger's flank and rear, to provide against a flank attack from the west. The low ground along the little brook made a continuous line unadvisable, and Kimball occupied the higher grounds northward toward the river, so that his brigades could mutually support one another as well as cover Ruger's flank. His barricades were of a slighter kind than our intrenchments on the front, and were necessarily more hastily constructed. As they were not continuous, and had no abattis, the Confederates expressed their regrets afterward that they had not known this, so as to modify their attack accordingly.[3] No doubt an attack in force by Hood's left would have changed the character of the battle; but it by no means follows that it would have been more success-

---

[1] O. R., xlv. part i. p. 351.

[2] *Id.*, p. 1172. The order was nearly identical in tenor with that which had sent the batteries to me. It was as follows, viz.: —

"HEADQUARTERS ARMY OF THE OHIO, FRANKLIN, TENN.,
November 30, 1864.

"GENERAL KIMBALL, —

"The Commanding General directs that you report with your command to Brig. Gen. J. D. Cox for position on the line to-day.

"Very respectfully,

"J. A. CAMPBELL,
Major and Assistant Adjutant General."

[3] See Hood's report, O. R., xlv. part i. p. 653, and especially General Bate's report, *Id.*, p. 743.

ful. In fact, it is doubtful if he could have changed the position of his army in time to have delivered an assault in that form before dark, and, by morning, we should have evacuated the town. We had rightly judged of his probable action and of the points of his heaviest assault, and had nothing to correct in the arrangement of our main line and in the distribution of its forces.

Kimball's division consisted of three brigades,[1] and he placed that of Grose on his left, immediately supporting the right of Ruger and the Pennsylvania battery. Kirby's brigade was in the centre, and Whitaker's on the right. As already indicated, the general line of this division was nearly north and south, facing to the west. General Grose's brigade was made up of seven small regiments,[2] and in his first line were the 84th Indiana, 30th Indiana, 75th Illinois, and 9th Indiana, numbering from left to right. The three first named were in continuous line, reaching to the bank of the little watercourse which here makes a large curve to the east. The 9th Indiana was on the north side of this hollow, and drawn back about a hundred and fifty yards. In Grose's second line were the 80th and 84th Illinois, the latter being so placed as to cover the ravine and low piece of ground between the 9th Indiana and the rest of the first line. The 77th Pennsylvania was put on duty as skirmishers covering the large curve from the Carter's Creek Turnpike northward. Kirby's brigade had the 21st Illinois, 38th Illinois, and 31st Indiana, with two companies of the 101st Ohio in the first line, supported by the 81st Indiana, 90th Ohio, and eight companies of the 101st Ohio.[3]

[1] O. R., xlv. part i. p. 177.   [2] *Id.*, p. 208.
[3] *Id.*, pp. 184, 1197.

General Whittaker's brigade, which was the right of the division, was stretched out northward toward the river, having the 35th Indiana, 21st Kentucky, 40th Ohio, and 45th Ohio in line, the 96th Illinois, 23d Kentucky, and the 41st Ohio in reserve.[1]

The placing of Kimball's division on our right completed the array in front of the village of Franklin. The troops at the breastworks were busy getting their noonday meal. The Columbia road had been crowded all the morning, double lines of wagons and of cannon hastening through the opening in the works at the centre. Wood's division of the Fourth Corps had marched across the new foot-bridge as soon as it was completed, and was placed on the commanding north bank of the river with the artillery of the Twenty-third Corps, to cover the final crossing of the army. The army wagons were laboriously pulling through the ford, and, after some careful experiments, were being cautiously driven over the bridges which had been meant only for the marching troops. General Schofield had industriously supervised all that had been done under his orders, and soon moved his own headquarters to the north side also. He reckoned that the wagon trains could all pass the river before sunset, and that the troops could be withdrawn soon after, if the enemy should make no attack.

There was now a period of rest and refreshment for the officers and men of the main line. Quiet followed the rattling of wheels and the clatter of arms that had made a continuous din in front of the Carter house all the morning. Our camp dinner over, the tents at my headquarters were struck, the baggage packed, and the wagons sent into the

[1] O. R., xlv. part i. p. 195.

town to fall in at the rear of the trains when the rest should be over the river. Our horses were fed and saddled, and the group of orderlies lounged on the grass by the roadside at the foot of the hill, while the officers were sitting in the veranda of the house, smoking or sleeping, as the mood took them. The day had proved to be a bright and warm one, a good sample of Indian summer weather coming after the first sharp frosts and snows of opening winter. The air was hazy, and, except an occasional straggler following his command in, nothing was to be seen between us and the Winstead Hill, two miles away. A distant cannon shot now and then told us that Wagner's division of the Fourth Corps, the rear guard, was checking the enemy's advance on the other side of the hill.

CHAPTER IV

THE REAR GUARD — WAGNER'S DIVISION.

Guarding the Trains — Collecting the Stragglers — Halt on Winstead Hill — Conditional Orders — Skirmishing with Forrest — Hood's Infantry press close — Will he turn the Position or attack ? — Wagner withdraws Lane's Brigade to Privet Knob — Conrad's farther in Rear — Opdycke's within our Works — Wagner's Message to Stanley — His Colloquy with Opdycke — His Orders to Lane and Conrad to Fight — The Sergeants to fix Bayonets — Ruger's Preparations — Confederates forming for the Attack.

WAGNER'S division of the Fourth Corps had well performed a laborious and difficult task in acting as rear guard. Two of its brigades, Bradley's (commanded by Colonel Conrad after General Bradley was wounded at Spring Hill) and Lane's, were on the flanks of the wagon trains, and marching with them as guard, in a way similar to that of the rest of the troops of the corps. Opdycke's brigade retired in line, halting and facing to the rear when they were too closely pressed by the enemy's cavalry.[1] The troublesome duty of getting forward the stragglers and footsore had of course fallen upon them. At eleven o'clock the division reached Winstead Hill,[2] and, by direction of General Stanley, Wagner deployed his whole command. Opdycke's brigade

---
[1] O. R., xlv. part i. p. 231.
[2] In Wagner's and Opdycke's reports this is called Stevens Hill, but I use the name which was then as now in common use in the vicinity.

was at the gap over which the turnpike runs, but also occupied the high point on the east, having there a section of Battery G, 1st Ohio Light Artillery. Lane's and Conrad's brigades extended the division line eastward toward the Lewisburg Turnpike. The division stacked arms, and the men were allowed to get their breakfast. At this time Whitaker's brigade of Kimball's division occupied the high ground west of the gap, with a section of Battery M, 4th U. S. Artillery, but was under orders to follow the rest of that division into the town when the road should be clear.[1]

It was not long before the enemy's advance guard appeared, and Wagner called his men to arms. He soon noticed, however, that Whitaker's brigade was marching toward Franklin, and having then no special orders to separate himself from the rest of the column, he directed his brigades to follow. Opdycke's brigade at the gap was again the rear guard,[2] and his statement is that he had been at rest about an hour and a half. Wagner, at the head of his division, had come within half a mile of the town,[3] as he says (he probably means our main line of works), and Opdycke's rear was just leaving the hill when the column was halted, and Wagner ordered its return to the heights they had just left.[4]

The occasion for this "right about face" is thus stated in his official report.[5] "I met a staff officer from General Stanley, with written orders, directing me to reoccupy the heights and hold them, unless too severely pressed." He is equally explicit as to what he did. "In obedience to this order," he says, "I returned to the position from which I had

[1] O. R., xlv. part i. pp. 195, 338.   [2] Id., p. 240.
[3] Id., p. 231.   [4] Id., p. 240.   [5] Id., p. 231.

just withdrawn my command, except that I now detached one regiment from Colonel Lane's brigade, and directed it to be placed on the heights to the right of the pike, from which General Whitaker had just withdrawn his brigade."[1] The reports of his brigade commanders corroborate him. Opdycke says, "When I reached the top of the hill I at once discovered heavy and parallel columns of infantry approaching rapidly. I was ordered off again after sending a number of shell and solid shot at the advancing enemy."[2]

Before following this second retrograde movement, it will be well to analyze more fully the circumstances under which the written order was sent to Wagner, which he interpreted as a command to return to the crest of Winstead Hill. We are fortunate in having the text of the order preserved.[3]

"HEADQUARTERS FOURTH ARMY CORPS, FRANKLIN, TENN.,
November 30, 1864, 11.30 A. M.

"BRIGADIER GENERAL WAGNER,
Commanding Second Division,—

"The general commanding directs that you hold the heights you now occupy until dark, unless too severely pressed; that you relieve Colonel Opdycke with one of your brigades, and leave his and the remaining brigade as a support; and that you cross the river to the north bank after dark, at which time the position you are to occupy will be pointed out to you.

"Very respectfully, your obedient servant,
"[J. S. FULLERTON,]
Assistant Adjutant General."

The understanding at headquarters of the Fourth Corps is given in a journal kept by Lieut. Colonel

[1] O. R., xlv. part i. p. 231. [2] *Id.*, p. 240. [3] *Id.*, p. 1174.

Fullerton, who apparently penned the order.[1] This has the following entry, viz.: "12 M., Colonel Opdycke reaches the high knoll two miles from Franklin; and General Wagner is ordered to hold him in this position and support him with his other two brigades until he is seriously threatened by a superior force of infantry."[2] General Stanley was at this time at General Schofield's headquarters in the village, at the house of Dr. Clift, near the public square, and it was about this hour of the day that Schofield determined to withdraw everything to the north bank of the river at dark, if the enemy should not attack. Schofield soon afterward moved his headquarters across the Harpeth, to the house of Mr. Alpheus Truett, and Stanley accompanied him.[3] Schofield had no wish to have any part of his army seriously engaged with the river at its back; but the crossing of the trains was work for the whole afternoon, and my task of covering this crossing would not be complete till all should be over. If Hood did not attack before dark, the cover of night would manifestly be a security for the safe transfer of the troops and artillery in the line.

The hostile force with which Opdycke had been skirmishing seems to have been Forrest's cavalry; for the first distinct mention of columns of infantry

---

[1] The brackets enclosing the name of Colonel Fullerton in the copy of the order indicate that the order is not the original which was handed to General Wagner, but is the official record of it kept at corps head quarters, and has no signature. I think it more probable that it was signed by Major Sinclair, also Assistant Adjutant General on Stanley's staff; for Colonel Fullerton in making the entry in his diary was evidently not aware of the fact that Opdycke's brigade was changed from the advanced position to that of support by the order itself.

[2] O. R., xlv. part i. p. 149.

[3] Mr. Truett is still living (1896). His house is on the east side of the Nashville Turnpike, about half a mile north of the river.

is when our rear guard went back to the top of Winstead Hill. It seemed most probable that Hood would use the same strategy as at Columbia, and force us to make a farther retreat by crossing the Harpeth either above or below us. The correspondence between Thomas and Schofield is based on this theory, and none of us were quick to believe that a *coup de main* would be attempted.

The order to Wagner which has been quoted was evidently written on the supposition that he was still in position on Winstead Hill. It was "the heights you now occupy," and the really material part of the order was the direction to retire at nightfall, and sooner if "too severely pressed." Except for Wagner's information, there was no need to hurry the issue of the order for the general movement to the north bank, and it was sent to me at the time Schofield's headquarters were taken across the river. My own report states that I received it at two o'clock.[1] The order, as issued by General Schofield, does not seem to have been preserved, and was probably delivered orally by a staff officer. That which was issued by General Stanley to Kimball's and Wagner's divisions of the Fourth Corps was in writing. It directed those officers to "commence withdrawing your command at dark. General Kimball will take the foot-bridge, and General Wagner the railroad bridge. General Kimball will not withdraw his pickets until 12 o'clock to-night."[2] It was signed by Major Sinclair, the assistant adjutant general before mentioned, and was, of course, in conformity with the order from General Schofield, which went to Stanley and myself alike.

But when Opdycke reached the gap in Winstead

[1] O. R., xlv. part i. p. 352.   [2] *Id.*, p. 1174.

The Rear Guard — Wagner's Division 69

Hill in obedience to the order of Wagner to return there, a new condition of things was revealed. The enemy's infantry was now plainly seen coming on rapidly in parallel columns. The section of Marshall's battery which was still with the division opened on the columns with shot and shell. The two turnpikes from Columbia and Lewisburg draw near each other as they approach Franklin, and Wagner, finding that the enemy's column on the Lewisburg road was turning his left flank, promptly determined to withdraw again, and sent an officer of his staff to notify General Stanley of the movement.[1]

He now followed the directions of the order he had received by giving to Lane's brigade the advanced position, putting Opdycke and Conrad in support. The section of artillery was with Lane on Privet Knob, which afterward became the place of Cheatham's headquarters in the battle. It has already been described as a stony hill on the west of the Columbia Turnpike, half way between Winstead Hill and the Carter house, and the most considerable elevation in the general plain surrounded by the higher hills like the Winstead.

The officer sent by Wagner to notify General Stanley of his retiring from Winstead Hill is said to have been Captain Whitesides, his acting adjutant general, and a statement has been published purporting to be his, in which he speaks of carrying a message to General Stanley at about 2.30 P. M., reporting the strong objections made by Colonel Lane to keeping his brigade and Conrad's out any longer. The statement contains the additional fact that he found General Stanley at Schofield's headquarters in

[1] O. R., xlv. part i. p. 231.

Dr. Clift's house in the village, and that no orders were sent back to Wagner.[1] This helps to make it clear that the message was the one Wagner says in his report was sent by him immediately after retiring from Winstead Hill, though the time of day is probably an error of recollection, which is corrected by Colonel Fullerton's note of its reception in his journal. No subsequent message is mentioned by Wagner as having been sent by him before the battle, and any sent as late as half past two would not have found Generals Schofield and Stanley at Dr. Clift's in the village. Colonel Fullerton's journal reads thus:—

"1 P. M. General Wagner reports two large columns of the enemy's infantry approaching Colonel Opdycke's position, moving on the Lewisburg and Columbia pikes, and, as he cannot successfully resist the forces, he is moving his division within the bridge-head constructed by General Cox around the town of Franklin."[2]

An approval of this by his superiors was all the answer the message required, and whether it were given tacitly by the mere acknowledgment of its receipt or by a more formally expressed assent is a matter of no moment. Wagner was acting within the spirit of the instructions given him, and only needed to complete his movement as he had begun it. It is plain, however, that in some manner instructions were given to him that upon coming within my lines — the "bridge-head around the town," as Colonel Fullerton calls the works — he was to place his command by my direction where it would be conveniently in reserve, and could be called upon in case of need. The official evidence of this is clear.

[1] See Van Horne's Life of General Thomas, pp. 289, 291.
[2] O. R., xlv. part i. p. 149.

In his official report (dated December 2d),[1] General Wagner states that, when the command retired from Stevens (Winstead) Hill, "he directed Colonel Opdycke to form in the rear of Carter's house, to the right in rear of the main line of works, to act as a reserve."

In my own brief preliminary report (dated December 2d),[2] I said that General Wagner reported to me, giving the situation of his brigades, Opdycke's being already in reserve within the lines. That he further informed me "that he was already under orders to keep out the two other brigades till the enemy should make an advance in line in force, when he was to retire skirmishing, and become a reserve for the line established by me." In my full official report (dated January 10),[3] the same facts are narrated with more detail, and accompanied by my own directions to Wagner as to position and duty when he should come within the lines.

General Schofield's report (dated December 31) says of the heavy loss in Wagner's two brigades, that it arose "from their remaining in front of the line after their proper duty as outposts had been accomplished, and after they should have taken their positions in reserve."[4]

The concurrence of all these leaves no room for dispute that Wagner was to come within the lines and act as reserve, and that he understood his orders accordingly. He in fact came in person within the intrenchments, at the head of Opdycke's brigade, and this was placed in close column of regiments

---

[1] O. R., xlv. part i. p. 232.
[2] *Id.*, p. 348.
[3] *Id.*, p. 352. See Appendix B.
[4] *Id.*, p. 344. See Appendix A.

on the right of the turnpike, about two hundred yards north of the Carter house.[1]

In my account of the battle, written in 1882, and published in another form,[2] I have stated that "the commandant upon the line was notified by General Schofield that Wagner's orders directed him to remain in observation only till Hood should show a disposition to advance in force, and then to retire within the lines to Opdycke's position, and act as a general reserve. Wagner, on being shown the note conveying this notice, said that such were his orders." The note referred to does not seem to have been preserved, and the statement has only the weight which would be given it by clear recollection. It was corroborated by much earlier statements to the same effect contained in my correspondence, as well as by the assent of Generals Schofield and Stanley. Being clearly in accord with the evidence already presented from the Official Records, it adds to the story an additional authentic detail among the circumstances attending the situation between two and three o'clock in the afternoon.

We shall get some light upon Wagner's condition of mind and purpose, by going back to consider more carefully the steps in the movement from the Winstead Hill about one o'clock. The written order from Stanley's headquarters directed that Opdycke's regiment should be relieved from the duty of extreme rear guard, which it had been gallantly performing all the morning. This was an unusual interference with the discretion of a division commander having his whole division in hand, and indicates some chafing between him and his subordinates

[1] O. R., xlv. part i. pp. 240, 352.
[2] The March to the Sea, Franklin and Nashville, p. 86.

which made the action of the corps commander necessary.

Wagner, having put Lane's brigade with the section of artillery in position on Privet Knob,[1] Opdycke and Conrad were marched to the point about half a mile in front of our main line. The official reports are silent as to a discussion which occurred here, but it is narrated in a published account of the campaign written by Colonel Opdycke.[2]

"On the way in from Stevens (Winstead) Hill, Opdycke was ordered by Wagner into line with the two brigades; but having in a former campaign become familiar with the military features of the locality, he thought the position so extremely faulty that he objected, and so was allowed to go into reserve on the rear slope of Carter's Hill, about two hundred yards from the main line of earthworks. Wagner then said to him, ' Now, Opdycke, fight when and where you think best: I may not see you again.' "[3]

It must be kept in mind that Privet Knob, where Lane was halted, was midway from Winstead Hill to our lines near the Carter house, the distance being about a mile each way. Conrad's brigade was put in position where Lane finally joined him, the place being variously stated as from three hundred or four hundred yards to half a mile from the lines. The larger distance is that which I gave in my official report, and this was written after I re-examined the field in our pursuit of Hood three weeks later.[4]

[1] *Ante*, p. 69.
[2] The New York Times, September 10, 1882.
[3] In private correspondence with me, in February, 1876, General Opdycke used still more vigorous terms to express his unwillingness to halt in the advanced position.
[4] O. R., xlv. part i. p. 352. This is conclusively corroborated by

As Wagner's purpose seems to have been to withdraw by alternately passing his brigades, it was probable that he would put Conrad's in support of Lane's at about half the distance from Lane's to the lines within which Opdycke's was marching to take position. This arrangement in echelon was not a bad one if the brigades were not allowed to become involved in a serious fight with superior forces. Lane should have left his hill in time to pass Conrad and join Opdycke in reserve, and Conrad should have followed, so as also to be wholly out of the way when the advancing enemy should be within range of fire from my lines. Wagner's preliminary arrangements all indicated this purpose. It was one which any competent division commander handling a rear guard is presumed to be familiar with, and for which he should need neither instruction nor suggestion from his corps commander. The preliminary steps had been well taken under his actual orders, which have been fully detailed. There was no change in his orders. At headquarters of the Fourth Corps and of the army, we have seen that he was supposed to be in the act of thus deliberately and progressively retiring.[1] We have to inquire how this intelligent plan was interrupted in execution, and became, instead, a nearly irretrievable blunder.[2]

Colonel Carter, the owner of the farm, who says the surveyed distance from his house to the first rise of ground on the turnpike southward, is 160 rods. This first rise of ground was the position of Conrad and Lane. See p. 46, *ante*.

[1] See the quotation from Colonel Fullerton's journal, *ante*, p. 70.

[2] In the newspaper discussion in which, at one time, survivors of this campaign largely indulged, it was sometimes argued that General Schofield must have given personal orders that Wagner's two brigades should fight in front of the line which he had so carefully planned as the only one on which to resist Hood. It is suggested also in Van Horne's Life of Thomas, pp. 289-291. Such a contention is sufficiently

Colonel Lane tells us in his official report that at two o'clock he sent word to General Wagner that the enemy was advancing in force and was about to envelop his flanks.[1] "With my skirmish line, and a section of artillery posted on Stone Hill" (Privet Knob), he adds, "I retarded the advancing column until I received orders, and withdrew my command to a position one third of a mile in advance of the main line of works, on the right of the Third Brigade (Conrad's)." The report of Captain Marshall of Battery G, 1st Ohio Artillery, fully corroborates Colonel Lane. He tells us that at 1 P.M. he moved Lieutenant Mitchell's centre section of the battery out on the Columbia Turnpike, and upon a hill near the skirmish line (Lane's on Privet Knob).[2] Here he opened on the enemy, who was found advancing in strong force. At 2.30 P.M. he withdrew the section, in compliance with orders, "into the turnpike within our first line of works, and continued firing." He thus accompanied Lane's brigade in its retreat to Conrad's line, which he calls the "first line of works" as distinguished from "our main line," which he mentions in the next sentence.[3]

Wagner, in his official report, agrees with the account given by Lane. The latter was halted on Conrad's right, instead of marching past and entering the intrenched position occupied by the Twenty-

answered when it is plainly stated. It will be seen that the result of my compilation of official evidence is to relieve the Fourth Corps headquarters also of any imputation of error in the matter.

[1] O. R., xlv. part i. p. 256.

[2] *Id.*, p. 331.

[3] The Confederate accounts also speak of the line occupied by Conrad and Lane when they were struck by the enemy as our first line. This is apt to lead to confusion, for in the fight about the Carter house it will by and by appear that there were two lines there in the latter part of the engagement.

third Corps. Conrad's arrangement of his troops was not wholly on the left of the turnpike.[1] He placed one regiment on the right, leaving the road wholly open. The other regiments were finally placed in single line. Their order from right to left was as follows, viz.: 15th Missouri, 79th Illinois, 51st Illinois, 42d Illinois, 64th Ohio, and 65th Ohio. When Lane came back to this line, between half past two and three,[2] the section of Marshall's battery occupied the open road in the interval between Conrad's right regiments.

Colonel Lane does not give, in his report, the order of his regiments in line, and we cannot supply the omission by the separate regimental reports, for none were made except by the 97th Ohio, his own regiment. The regiments in the brigade were, besides the one named, the 26th Ohio, 28th Kentucky, 40th Indiana, 57th Indiana, and 100th Illinois, and it is understood that, like Conrad's, they were deployed in single line.[3] Each brigade had its outer flank well retired, so that the whole formation was almost wedge-shaped.[4] As to his orders from General Wagner, Lane is sufficiently explicit. He says, "I here received orders to give battle to the enemy,

---

[1] O. R., xlv. part i. p. 270.

[2] Lane sent to Wagner for orders, as we have seen, at two o'clock, and waited to receive an answer at Privet Knob before beginning his movement. Wagner was then just bringing Opdycke's brigade within my lines, and in going and returning the messenger must travel two miles. The movement then began, and the infantry marched half a mile to Conrad's position. All this would take more than half an hour.

[3] O. R., xlv. part i. p. 255.

[4] The section of artillery took position in the road far enough north of the low summit in front of my lines to make the curve of the ground partly cover the guns. Conrad formed on this position, a hundred yards or more on the hither side of the crest at the road. See map, p. 45, *ante*.

and, if able, drive him off; if overpowered, to check him as long as possible, and then retire to the main line of works."

In his visit to my headquarters at the Carter house, when he came in with Opdycke, Wagner had himself seen the character of the lines. The intrenchments, the artillery at the embrasures, the retrenchment covering the road, the abattis in front, were all examined, and made the subject of conversation by him. His experience in the construction of such lines and their purpose ran through the whole of the Atlanta campaign, and he not only knew that the lines were planned to meet any serious attack the enemy might make, but he had just reported to headquarters of the corps to which he belonged that he was moving his division within these lines. As, in addition to this, his official report makes no suggestion of any change in his orders from any source, it is conclusive that the change in his conduct must have been spontaneous with him.

Conrad's arrangement of his troops so as to leave the road open, and his placing a regiment on the right of it, plainly looked like a temporary halt, with the expectation of covering the march of Lane's brigade down the turnpike. There is an indication, too, that he did not at once deploy his brigade in single line, but placed one regiment on each side of the road and the others in support. The barricade which his men made, imperfect as it all was when compared with our more carefully constructed lines, was yet a stronger defence close to the turnpike than it was on the outer flanks, showing that this outer part was more hastily thrown up.[1]

---

[1] Colonel Carter states that he personally levelled down these barricades, and that where Lane's brigade stood, on the west of the road,

When the enemy began to deploy on the hither side of Winstead Hill, Conrad sent to inquire of Wagner whether it was expected that he should hold this line; but, as he reports, just as his officer was starting, Wagner rejoined him in person, and not only gave orders to hold the line as long as possible, but directed him to have the sergeants fix their bayonets, and keep the men to their places.[1] "I accordingly gave the same instructions to my regimental commanders," Conrad adds, "and, believing an attack would soon be made on my line, I ordered them to build a line of works in front of their regiments respectively." How much time there was for this attempt at intrenching we cannot accurately tell, but it is doubtful if more than half an hour elapsed when the enemy advanced to the attack. Conrad says it was while his men "were very busily engaged in throwing up the work."[2] As to Lane's brigade, the only evidence is the report of Lieut. Colonel Barnes, who says they "had but fairly begun to throw up a temporary work."[3]

All accounts show that there was neither time nor means to make a solid intrenchment at this outpost. The regiments had marched without intrenching tools, and it is said that, finding a wagon broken down by the roadside, in which were some shovels and pickaxes, they had helped themselves to these.[4] The army and corps commanders were ignorant that they had any intrenching tools at all.[5]

they were not continuous, but were in the shape of disconnected "half moons," and that all were of the hasty kind made by throwing fence rails together and piling a little earth behind them.

[1] O. R., xlv. part i. p. 270.
[2] *Ibid.*   [3] *Id.*, p. 265.   [4] *Id.*, p. 280.
[5] Major Atwater, commanding 42d Illinois, says, "when the enemy came upon us we had a very poor line of works." *Id.*, p. 276. Cap-

It would seem, therefore, that General Wagner, chafing at the urgency, akin to insubordination, with which Opdycke had wisely opposed any stop in the continuous retirement of the division after passing Privet Knob, and excited by the rapid approach of a crisis in the stirring events of the day, gave way to an impulse to fight the whole army of Hood upon the line of mere outposts. Such impulses, unfortunately, are not uncommon in officers who are brave enough, but who lack the power of calm self-control under fire.[1] He was thus led to issue the order which involved the two brigades in terrible consequences.[2] In his own report, Wagner said that

tain Atwater, commanding 51st Illinois, says his men "threw up hasty works." *Id.*, p. 278. Colonel Buckner, of the 79th Illinois, says the "works were constructed in a short time," and "it was not long till the skirmishers were engaged in our front." *Id.*, p. 280. Lieutenant Colonel Brown, commanding 64th Ohio, says that "the men with a few spades, threw up a bank, which, in consequence of no timber, was very low." *Id.*, p. 284. Lieutenant Colonel Barnes, commanding 97th Ohio (Lane's brigade), says, "we had but fairly begun to throw up a temporary work, with the very limited means at our disposal," when the enemy attacked. *Id.*, p. 265. All of these except the last are in reports from Conrad's brigade, and as we have seen that their barricades were much better than Lane's, it is fair to say that the latter, the disconnected "half-moons," were scarce worth mentioning.

[1] A noteworthy example of this impulse to conduct a battle by "feeding the pickets," is that attributed by Kinglake to General Pennefather at the opening of the battle of Inkerman in the Crimean War. Kinglake's Invasion of the Crimea, vol. iii. pp. 101, 154 (Harpers' edition).

[2] Conrad's report was made on December 1st. The division reached Nashville about noon of that day, and though he had been fighting and marching two days and nights without rest, the report was written before he slept. Wagner's was made on the 2d, though his subordinates, Opdycke and Lane, did not make theirs till the 5th and 7th respectively. Conrad's reporting that he was ordered to fight, even by making the file closers use bayonets, was equivalent to "charges" against Wagner, and the latter hastened his own report to reply to it, not waiting to receive those of his other brigade commanders. I shall have occasion to notice the sequel of this. *Post*, chap. xvii.

Conrad and Lane were "directed to hold their position long enough to develop the force of the enemy, but not to attempt to fight if threatened by the enemy in too strong a force."[1] He thus puts himself in accord with the orders which he had received from his superiors, and makes by implication a denial of the truth of Conrad's statement.

From the breastworks at the Carter house the enemy could be seen as a dark line coming diagonally down Winstead Hill, blotting out the white streak of the turnpike. Two miles is too long a distance, however, to see more than a very general outline of a moving column, livened here and there by the glint of light flashing from a gun barrel. At the foot of the hill the ground was not only lower than it was at one or two points nearer to us, but there was an open wood or grove with forest trees enough to hide the deployment when they began to prepare for the attack. For an army of thirty thousand men it takes two or three hours at least to pass from column into line, properly aligned, with supports also in position, artillery in the proper intervals, and all properly "dressed up" with accuracy; because this accuracy is more necessary than on parade, when the event of a great battle hangs on the unity and vigor with which an assault is delivered.

After my interview with Wagner, when Opdycke's brigade came within the lines, I had made a final inspection of our preparation. I had visited General Ruger, who, knowing that I expected to be personally at my headquarters at the Carter house, had rightly judged that the next critical point would probably be at the Carter's Creek road, where the Bostick place might give the enemy a commanding

[1] O. R., xlv. part i. p. 231.

position for artillery. He therefore fixed his personal position on the line in rear of his right brigade. The recurved part of his line was in danger of enfilading fire, and he had constructed it as a *crémaillère*,[1] the indentations in it having much of the effect of traverses in protecting the men. In at least one of the regiments regular traverses were also built between the companies. The locust grove near the Carter house, and the orchard beyond, had both disappeared under the demand for material to make abattis, and the clear range for infantry fire was unobstructed for half a mile. General Reilly, in command of my own division, made his headquarters at the cotton-gin, where the salient occupied by the battery was not only a point likely to be aimed at in an assault, but was also prominent enough to give a view along the front if the smoke should permit.

The troops in the line were everywhere agog with the news of the imposing formation of the enemy in full view of Wagner's advanced brigades. From officers coming to me with reports from our left, I learned that the hostile array could be well seen from the knoll in Stiles's brigade, and I determined to go there for a brief visit, to judge for myself of Hood's organization and the points at which his attack would be aimed. It was now three o'clock, and as we were only three weeks from the winter solstice, the sun would set at five o'clock. Hood must attack soon or not at all. Wagner had come

---

[1] In the official map of the field made by Captain Twining, the Chief Engineer of the Twenty-third Corps (see map, p. 45), part of General Ruger's line is drawn as having traverses inside a straight breastwork. General Ruger's official report, however, says that it was "a broken line." Colonel Spaulding, 23d Michigan, says he built "traverses on the flanks of each company." O. R., xlv. part i. pp. 365, 386.

in from his two brigades, and, meeting him in the road in front of the Carter house, he confirmed the information that the enemy was probably forming for an assault. I reminded him of his orders not to leave his brigades out too long, and warned him of the dangers that would come from a hurried retreat. I then rode off to the left.

General Schofield's staff had kept up frequent communication between us during the day. Colonel Schofield had just inspected the position of the additional batteries which Captain Bridges had put in position. Colonel Wherry, the Chief of Staff, had just carried the latest news from the front. On the report of the deployment of the enemy, General Schofield had gone to Fort Granger, from the parapet of which, much higher than any point in my line, he had a complete view of the whole field.

At a quarter past three General Wilson found that the enemy's cavalry were pushing back our own mounted outposts, and were making efforts to force a crossing of the Harpeth three miles up stream, at Hughes's Ford.[1] There had been brisk skirmishing for an hour before, and Croxton's cavalry brigade had retired to the north bank of the river at McGavock's Ford, about a mile below Hughes's. The river, so far up, was reported by the inhabitants to be fordable in many places, and as soon as the advance of the enemy was developed, Wilson concentrated Hatch's and Johnson's divisions to resist it. Thus the ball opened nearly simultaneously on the Columbia Turnpike by the infantry assault upon Wagner's brigades, and at the up-river fords by Forrest's attack upon our cavalry. All this will grow clearer as we now trace the enemy's advance from Spring Hill.

[1] O. R., xlv. part i. pp. 559, 560, 1184.

MAP OF
A PORTION OF THE
BATTLE-FIELD
OF
RANKLIN, TENN.,
SHOWING THE POSITION
OF
STEWART'S CORPS.
COMPILED FROM SURVEYS
BY
APHICAL CORPS, ARMY OF TENN.

Scale
0   1000   2000   3000   4000 Feet

W. F. Foster, Maj. & Sen'r Eng'r
Stewarts Corps.

——— Union
——— Confederate

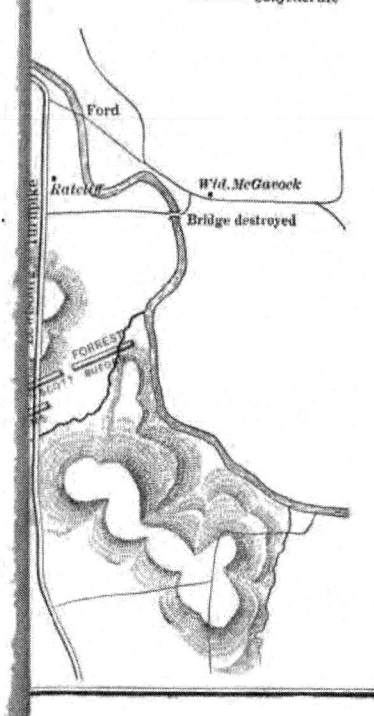

CHAPTER V

THE CONFEDERATE ARRAY

Hood at Spring Hill — Discovers Schofield's Escape — Cavalry in Pursuit — Infantry hastening after — He decides to Assault — His Cavalry Positions — The Infantry — Deployment and Formation — Stewart's Corps on Hood's Right — Cheatham's in Centre and Left — Part of Lee's in Reserve — Artillery in Intervals — Chalmers's Cavalry on extreme Left — Hood's Headquarters.

It was not till daylight of the 30th November that Hood learned at Spring Hill that Schofield's little army had evaded him. Forrest was dispatched in pursuit with Buford's and Jackson's divisions of cavalry; but Chalmers was sent with his division to explore the roads leading westward into the Carter's Creek Turnpike, Hood being unable to persuade himself that all Schofield's troops and trains could have moved by the direct road to Franklin in the night.[1] He had suspended operations late in the evening of the 29th, in the belief that Forrest blocked the way to Franklin, and had persuaded himself that Schofield must surrender in the morning.[2] Chalmers fully expected to find that part, at least, of our column had been forced to leave the Columbia and Franklin

---

[1] O. R., xlv. part i. p. 764.

[2] See an important paper by General Cheatham, and the accompanying documents, in Southern Bivouac for April, 1885. Chalmers's report for his cavalry division is not preserved in the official archives, but a copy found among his papers is published in the Official Records, and is no doubt authentic. *Id.*, p. 763.

Turnpike by a detour to the west, and was amazed to find that our whole army had passed up the direct road to Franklin, within a very short distance of their infantry lines.[1] He continued his route, however, in accordance with his orders, and on reaching the Carter's Creek Turnpike he turned northward upon it toward Franklin, although it was demonstrated that no part of Schofield's forces had travelled that road.

Hood's infantry at Spring Hill consisted of Stewart's and Cheatham's corps, with Johnson's division of Lee's. He had with him also two six-gun batteries, — Guibor's and Presstman's. These were put in march for Franklin as soon as Forrest was fairly off. The cavalry had run short of ammunition in yesterday's fight, and as the ordnance trains were with the other wagons at Columbia, Forrest had to borrow some ammunition of the infantry. His men were really a mounted infantry, and were armed with guns of the same calibre with some of the line troops.

S. D. Lee's corps at Columbia, reduced by the absence of Johnson's division, consisted of Clayton's and Stevenson's divisions; but he had with him nearly all the artillery, and all the wagon trains. He became aware that we had left his front about half past two in the morning; but he had double the distance to march that the other corps had, and although he began his movement before day, he reached Franklin after the attack had begun, and too late to take part in it.[2]

---

[1] O. R., xlv. part i. p. 764.

[2] His report says that he reached Franklin about four o'clock, as Hood was about to attack; but he apparently had ridden forward in person, for Clayton says they found the bloody engagement begun. O. R., xlv. part i. pp. 687, 697.

The bold front shown by Wagner's division at Winstead Hill had forced Hood to deploy, for his failure at Spring Hill had not encouraged him to attack with the head of column. He tells us that, having gathered from the despatches he had captured at Spring Hill that Thomas was intending to hold Franklin in force by concentrating there, he knew it was all-important to attack Schofield before he could make himself strong.[1] It was undoubtedly wise for him to force the fighting with Schofield if he could catch him at a disadvantage, and crush him if possible before he could unite with Thomas's other troops, or be greatly reinforced at Franklin. At any rate, he determined, as he says, not to delay for flanking movements, but to attack at once in front.

He had followed up our column so closely that he might well assume that we had not had time to fortify, though his experience in the Atlanta campaign ought to have taught him that it did not take us long to throw up a formidable line of breastworks when we had deliberately chosen a position for defence. His exasperation at what he regarded as a hair's breadth escape on our part from the toils in which he thought he had encompassed us at Spring Hill had probably clouded his judgment. He blamed some of his subordinates for the hesitation which he seems himself to have been responsible for, and now, in an excitement which led him astray, he determined to risk everything upon a desperate assault.

Forrest's reputation as a "raider" was so great that it was apt to be assumed on the national side that he never confined himself to the strictly auxiliary work of a cavalry column accompanying an army. In this campaign, both on the advance and the retreat, he

[1] O. R., xlv. part i. p. 653.

proved that he could cover the movements of the infantry as brilliantly as he performed his independent raids. He had admirably assisted Hood as a flanking force all day on Tuesday, and now, on Wednesday the 30th, with his two divisions, he opened the way for the infantry, harassing our rear guard, and making dashes at the trains, sticking closely to his work of helping Hood forward, and with no thought of distant expeditions.

When, about one o'clock, Wagner's withdrawal of his division left the way over Winstead Hill open, Hood's infantry advanced at once by the Columbia and Lewisburg turnpikes, whilst Forrest's two cavalry divisions skirted the hill upon their right, leaving room for the deployment of the infantry.[1] Jackson's division was on Forrest's right, facing Hatch's division of our cavalry near Hughes's Ford, and Buford's on his left, near the Lewisburg Turnpike, was facing Croxton's brigade near McGavock's Ford. Both Confederate divisions were halted while the deployment of the infantry was going on, and Buford was ordered to advance close upon Stewart's right when the corps of the latter should go forward.

Chalmers had come up with the other cavalry division by way of the Carter's Creek Turnpike, and was directed to cover and extend the attack of Hood's left wing. By noon he was already in sight of Ruger's skirmishers, who were well in front of the right flank of the Twenty-third Corps line, accompanying the movement of Wagner's men as they withdrew to Privet Knob. It was on the report of Chalmers's appearance there, as we have seen, that I requested more troops to support and guard Ruger's flank.[2]

[1] Forrest's Report, O. R., xlv. part i. p. 754.
[2] *Ante*, pp. 59, 60.

Hood saw, as we did, that an attack by his right against our left promised most decisive results, if successful. If Forrest were able to drive Wilson back, our line of communications was most easily reached by that flank where the fords of the upper river were close to the line of the infantry advance, so that Stewart could aid Forrest as he went forward, and be aided in turn by every mile of vantage that the cavalry should gain.

The Confederate commander, therefore, not only kept two thirds of his cavalry under Forrest upon his right, but made Stewart take ground in that direction so far as to make room for Cleburne's division of Cheatham's corps between the Columbia Turnpike and Stewart's left. He thus deployed four divisions of infantry between the road which led to our centre and the Harpeth River on our left, whilst only two were deployed upon the west of the turnpike;[1] for Johnson's division of Lee's corps was kept in reserve until after the first assault, and Lee's other two divisions did not arrive on the field in time to take an active part in the engagement.

The order of battle on the Confederate side was an array of two brigades in front line and one in support in each division, with intervals for artillery between the divisions. The field of battle converged to a narrower front as the enemy approached our lines; for Reilly's single division of three brigades in our trenches was thus to bear the assault of four divisions, and Ruger's was to be attacked by two divisions, besides a subsequent onset by a division of Lee's corps. It will be seen, by and by, that the attack upon Ruger lapped over upon Kimball's left brigade, and that the advance against Kimball's centre and right was made

---

[1] O. R., xlv. part i. p. 708.

by Chalmers's cavalry; but Hood's orders contemplated guiding his infantry on the left by the Carter's Creek road, and the extension west of it was caused by the crowding of the lines as they converged upon our narrow front.

In his deployment Stewart had Loring's division on his right, consisting of Scott's, Featherston's, and Adams's brigades. Walthall's division was his centre, consisting of Quarles's, Shelley's, and Reynolds's brigades. French's division was Stewart's left, having at this time but two brigades, Cockrell's and Sears's, for Ector's was away acting as guard and escort for the pontoon train.[1]

Of Cheatham's corps, Cleburne's division, as I have already said, was east of the Columbia Turnpike, and took the road itself for its left guide. Its brigades were Polk's, Govan's, and Granbury's. Brown's division was on the west of the turnpike guiding its right by the road. It was made up of the brigades of Gordon, Strahl, Carter, and Gist. Bate's division was the left of the line, with orders to take ground to the west till it cleared Brown's division and Privet Knob,[2] then to advance guiding its left by the Carter's Creek Turnpike, in front of the Bostick house. His brigades were those of Finley, Jackson, and Smith.

Johnson's division of Lee's corps was held in reserve at the opening of the battle, until the rest of that corps arrived upon the field.[3] Its brigades were those of Brantly, Deas, Manigault, and Sharp. We shall have occasion to notice the arrangement of these brigades when we have to tell of their advance to the attack, and of those of all the divisions engaged. It is enough for the present to say that a space for artillery was left between the divisions of Stewart's corps, and that a

[1] O. R., xlv. part i. p. 708.   [2] *Id.*, p. 743.   [3] *Id.*, p. 681.

section of Guibor's battery was placed in each of these intervals, and between French and Cheatham.¹ Presstman's battery seems to have fought as a unit, being first placed on the slope of Privet Knob, where Cheatham had his headquarters, and later on the knoll at the Bostick place near the Carter's Creek Turnpike, and in Bate's division.

Chalmers's division of cavalry, which was ordered to advance beyond Bate's left flank, after meeting our skirmishers soon after midday, moved still farther to the left and front, so that when it came forward later to join in the general attack it came out of a hollow nearly at right angles to the line of General Bate's march. Whilst in this hollow it had been hidden from Bate's view as he advanced, and, as he was not aware of its exact position, he thought it had not come forward upon his flank in accordance with the plan of the engagement. It had really got into its concealed position before his deployment, and he does not seem to have learned that it took any part in the fight, for he reported that his left suffered for lack of the support the cavalry was to have given it.² We shall see more of this as the incidents of the battle develop.

Hood fixed his headquarters on the field at the old Neely house, not far from the foot of Winstead Hill, on the Columbia Turnpike. The position was not one from which he could view the engagement, and he had to rely upon the reports of his subordinates for his information of the progress of the battle. This, however, was of little moment, for the smoke of battle was soon to obscure everything and make personal observation of little use. He was a good deal exhausted in body by the labors and fatigues of the past two days, and in his maimed and crippled condition,

---

[1] O. R., xlv. part i. p. 708.   [2] Id., p. 743.

lacking a leg and an arm lost in former battles, he lay on the ground upon some blankets, a saddle supporting his shoulders and head, and so received the reports his corps commanders sent to him, and gave his orders in the desperate fight.[1]

[1] These particulars were given me by Major Sanders, of General French's staff, among his personal recollections of the battle.

CHAPTER VI

THE ASSAULT ON WAGNER'S OUTPOST

View from the Knoll on our Left — Skirmishing in Front — The Outpost trying to intrench — Confederate Advance — Colonel Capers's Description — Artillery opens on both Sides — Surgeon Hill's View from Fort Granger — Orders sent along our Line — To Opdycke in Reserve — Retreat of the Outpost — My Ride to the Centre — Momentary Break there — Reilly's Rally — Opdycke's Rush forward — Strickland's Rally — Meeting Stanley — The Din of Battle — Stanley wounded.

SUCH was the situation when I left my headquarters at the Carter house and rode along the lines of Reilly's division to the knoll in Stiles's brigade near our extreme left. It was known among the troops that the enemy was deploying, and officers and men were all upon the alert. From the parapet at the knoll, looking southward, a slight hollow ran diagonally up to the right toward the water-shed on which the Columbia road was built. This enabled me to look behind the rise of ground on which Wagner's two brigades were now deployed, and to see the gradually extending array of the Confederate army. It was evident that Hood was deploying, but it might be only for the purpose of encamping in line of battle just beyond the range of projectiles, as he had done at Columbia before beginning his flanking movement.

The day continued bright and warm; but the sun would set before five o'clock,[1] and it was already low

[1] See note at end of chap. x., *post*.

enough in the west to make the haze seem thicker under its slanting rays, which flashed from the weapons and the accoutrements of the enemy more than a mile away. Their line could be continuously traced from the Harpeth River on our left front till it was lost near Privet Knob, in front of our right centre, from which Cheatham's corps had driven Lane's brigade half an hour before, though our skirmishers were still holding the shoulder of the hill and making a lively resistance to the Confederate advance on that part of the field. Our cavalry on the river above us, at the fords, were considerably in advance of our left, and a brisk engagement between them and Forrest's men was already going on.

My attention was now attracted to the centre by the efforts of Lane's and Conrad's men to make some cover for themselves. They seemed to be digging hard to throw up a breastwork. I attributed it to the habit of our men to make some sort of shelter whenever halted, unless they certainly knew they were immediately to march, and I momently looked for the evidence that they were ordered to cease the useless labor and retire. From the Carter house I had not noticed any such work going on. Conrad's brigade seemed to stand by a fence, facing southward, and anything they had done to strengthen a barricade of rails had not been seen by me, though I had not had occasion to examine them closely, occupied as I was with my own duties.

Soon the long lines of Hood's army surged up out of the hollow in which they had formed, and were seen coming forward in splendid array. The sight was one to send a thrill through the heart, and those who saw it have never forgotten its martial magnificence. In our forest-clad and sparsely inhabited

Southern States it was a rare thing to have a battle-field on which the contending armies could be seen. We usually fought in tangled woods and thickets, where we knew the extent of the engagement only by the sound of the distant artillery and the crash of the musketry. Here, however, we could see the field and follow the movements of the contending hosts.

That it impressed our opponents, as it did ourselves, we know from many sources; but a vivid description, given in the report of Colonel Ellison Capers of the 24th South Carolina, enables us to see the field as it looked to an officer in the front line of Brown's division as it passed over the crest of Privet Knob, and marched against our works just in rear of the Carter house. Colonel Capers was destined himself to fall severely wounded in that desperate charge, and we can well understand that the beautiful landscape and the stirring pageant before him should be imprinted on his memory as it looked before he dashed into the sulphurous clouds of the battle storm.

Brown's division had come down from the Winstead Hill, where it had been deployed,[1] marching by the right flank of regiments until they reached the hollow at its foot. They then formed "forward into line," Gordon's and Gist's brigades in front line, supported by Strahl's and Carter's. "As we advanced," says Colonel Capers, "the force in front opened fire on us, and our line moved steadily on, the enemy retreating as we pressed forward.[2] Just before the charge was ordered, the brigade passed over an elevation from which we beheld the magnificent spectacle the battle-

---

[1] O. R., xlv. part i. p. 736.
[2] This was Lane's line of skirmishers on Privet Knob, before he joined Conrad in their last position. See Lane's Report, O. R., xlv. part i. p. 256.

field presented. Bands were playing, general and staff officers were riding in front of and between the lines, a hundred battle-flags were waving in the smoke of battle, and bursting shells were wreathing the air in great circles of smoke, while twenty thousand brave men were marching in perfect order against the foe."

The customs of military service discourage the indulgence in sentiment in military reports, but the results of this battle, and its terrible scenes, seem to have impressed Confederate soldiers with an unusual power, so that many of them speak with unwonted emotion in their reports, whilst many others are wholly silent, as at a great catastrophe of their cause. We are thus indebted to the extraordinary feelings the event excited for the vivid picture of the field as seen from the lines of the foe. I shall have occasion to borrow again from Colonel Capers's report, but must now return to the view from the left of our lines.

Although Stewart's corps had been formed in double lines, with two brigades in front and one in rear in each division, the rapidly narrowing field made it quickly necessary to reduce the front, so that by the time their lines could be distinguished from each other at our position, each division was a column of brigades, coming on at a quickstep with trailed arms. In the intervals between the divisions the artillery galloped forward, unlimbered, and fired till their lines passed to the front, then rushed forward again to repeat the manœuvre. The section of Marshall's battery which was with Wagner's men opened first on our side, and kept up a steady fire on the approaching enemy. Cockerill's battery in Fort Granger, north of the river, opened next with its long-range rifles, having a most commanding position, where they could sweep the front of our recurved left flank, or fire over the heads of our men in the lines. The batteries on both our flanks increased the fire as the Confederate columns came into range. Surgeon Hill of the 45th Ohio, the medical officer in charge of the field hospital of Kimball's division, was in Fort Granger when the battle opened. "Standing on this fort," he says, in a private letter, "where I could have an excellent view of the country, I could very well follow the flight of the heavy shells, and see their explosion among the enemy. The closing up of the broken ranks with such well directed and determined progress was most wonderful and commendable, even in an enemy. The artillery duel was most terrifically beautiful and grand, and our guns must have wrought immense damage." It was from this commanding position, it will be remembered, that General Schofield was directing his little army.

Uneasy that the two brigades in front were not already retiring, I sent an aid (Lieutenant Coughlan) down the line to warn the troops at the centre to withhold their fire till Wagner's men should get in, and to direct Opdycke to be ready to charge with his brigade if any break should occur. If Wagner should be at the lines near the Carter house, the order relating to his command would, of course, go to him; but if he were not found, the aid-de-camp was to deliver the order directly to Opdycke, and this was what he did.[1] I saw Mitchell retire his section of guns from Conrad's line, and come leisurely within the works on a slow trot. He had wound up his firing with canis-

[1] In his official report, dated December 5, Colonel Opdycke says, "General Cox sent me a request to have my brigade ready." O. R., xlv. part i. p. 240.

ter as the enemy drew near,[1] and had lost two men killed and five wounded in this exposed service with the outpost. Now surely the outlying brigades must begin their retreat! No, they did not, but they opened a rapid musketry fire in hopeless contest with the overwhelming masses of the enemy. For a little while they checked Cheatham's advance, but he outflanked them far on left and on right. Still they fought till their foes dashed over the frail barricades and were amongst them, and then they broke to the rear in confusion.

When I saw that the two brigades were committed to a hand to hand fight, I sent a second aid (Captain Tracy), with reiterated orders for Opdycke's brigade. The two officers were the only ones who had accompanied me on my ride to the left. I waited alone a few moments till I saw the break of the unfortunate outpost, then, after a final word with Henderson and Stiles, I mounted and galloped back toward the centre. I had meant to time my ride so that I should reach the centre before Wagner's men, but was delayed a little by an incident not unlikely to occur upon a battle-field. The artillery fire on our left and in Fort Granger had opened rapidly when the enemy's lines rose into view, and, as I passed behind the line (already firing) near the Lewisburg Turnpike, one of the enemy's shells, bursting close by, had frightened the horses attached to a gun limber, and the runaway team of six animals dashed by in the drifting smoke. This with the din of the discharges so scared my horse that he commenced rearing and plunging violently. He kept this up till, finding that I was making little

---

[1] See Captain Marshall's Report, O. R., xlv. part i. p. 331. Also report of Captain Bridges, Chief of Artillery, Fourth Corps, *Id.*, p. 321.

headway, I dismounted, and, holding him by the head, succeeded in quieting him by rubbing his nose and ears. Remounting, I went on my way; but a half-minute perhaps had been lost. When I came near the cotton-gin in Reilly's line, his reserve was just rising to the charge.[1] Seeing that these troops were going gallantly forward, I continued on toward the Columbia Turnpike. A glance showed that a break in the line had occurred there, and a crowd of men in the road was just passing out of sight among the nearest houses of the village. Between me and the Carter house a good many fugitives were still running to the rear, and some Confederates were advancing; but at that moment Opdycke's left regiments deployed from column into line across the road, between the crowd hastening into the village and the enemy who were trying to reform inside our works for a new advance. The Confederates were apparently hesitating to leave the cover of the breastworks and retrenchment, for the quick and stout rally on Reilly's line drove back their men who had got over our works west of the cotton-gin. The Carter house and outhouses prevented any charge by them in continuous line, and the splendid appearance of Opdycke's brigade deploying at double quick showed that the way was not open to them, though they had possession of our works to the extent of a brigade front. I could not see what was going on behind the Carter buildings, but the roar of the fight from that direction showed that only the left of Ruger's division had given back.

[1] See Captain Speed's description of the situation at this time, in his paper before referred to. Ohio Loyal Legion Papers, vol. iii. p. 44. Captain Speed was then adjutant of the 12th Kentucky Infantry in Reilly's second line, and speaks as an eyewitness. He is nephew of Mr. Speed, Attorney General of the United States, and is now the clerk of the United States Circuit and District Courts at Louisville, Ky.

Strickland's men were holding the buildings themselves, and a fierce fight was going on about them.

Changing my direction toward Opdycke's brigade, I passed the flank of his advancing line, and on the turnpike, urging his brave men to redoubled exertion, I joined Colonel Opdycke himself, and with him General Stanley and Captain Tracy. The latter was the aid-de-camp whom I had sent with the last order for Opdycke. Stanley,[1] as we learn from himself, had come to the front only a moment before, and just as Opdycke's men were rising to their feet and preparing to deploy. There was no time for conference or questioning. Every officer was fully employed rallying the disordered lines and cheering forward the advancing brigade. The scene was a startling one. The enemy filled the spaces about the Carter house, and were trying to form upon the turnpike in front of the buildings. They had possession of the cannon on both sides of the road, and sought to turn them upon us. They held in their hands a number of our men prisoners, but had not time to send them all to the rear.[2] As Opdycke's line charged forward, the roar of musketry on right and left and front was deafening, so deafening as to produce the effect of dumb show. The men looked as if breasting a furious gale with strained muscles and set teeth. Hood, who had ordered Cheatham and Stewart " to drive the enemy

---

[1] O. R., xlv. part i. p. 116. See Appendix D.

[2] Long afterward, an intelligent soldier of the 100th Ohio told me in graphic language how he was captured at the breastworks close to the cannon on the left of the road. He had stayed in his place till he was run over by our men retreating from the front, and the enemy followed so close that he could not get away. He saw the efforts to turn the guns so as to rake Reilly's line, but while they were looking for primers our second line and reserve were upon them and he was again free.

from his position into the river at all hazards," now heard the fearful din at his headquarters a mile and a half away, and his narrative shows how even he was impressed by it. "At this moment," he says, "resounded a concentrated roar of musketry, which recalled to me some of the deadliest struggles in Virginia, and which now proclaimed that the possession of Nashville was once more dependent upon the fortunes of war. The conflict continued to rage with intense fury."[1]

The distance from Reilly's reserve line to Opdycke's had been about a hundred and fifty yards.[2] I had passed just in rear of Rousseau's and White's Kentuckians and the East Tennesseans as they sprang forward, and now Opdycke's brigade was coming forward on their flank. The mêlée about the Carter house was desperate but short. Opdycke was in the thickest of it, and, after he had emptied his revolver at the enemy, he used it clubbed till it was broken. Stanley rode close in rear of the line, hat in hand, cheering them on. The Confederates were driven back over the works, leaving in our hands their dead and wounded, and a goodly number of other prisoners.

Opdycke's men now held the retrenchment which crossed the turnpike just beyond the Carter house;[3] his left regiment (44th Illinois) under Lieut. Colonel Russell, was on the left of the road joining the right of Reilly's command, and occupying the ground where the guns of the Kentucky battery which were retaken had been.[4] General Stanley now asked me to look at a wound he had received, and I saw that a ball had

[1] Advance and Retreat, pp. 293, 294.
[2] Opdycke's Report, O. R., xlv. part i. p. 240.
[3] *Ante*, p. 56.
[4] Reilly's Report, O. R., xlv. part i. p. 412. Russell's Report, *Id.*, p. 246.

raked across his neck, passing over his shoulder diagonally. The holes in his coat showed the place of the wound, and that it was not very deep, but, as the exit of the ball was close to the spine, I urged him to have surgical attention to it at once. We noticed also that his horse was disabled, and I dismounted my aid, Captain Tracy, who was riding one of my horses, and gave it to Stanley, who rode away toward the town. When we parted we were in the middle of the turnpike directly in front of the door-yard of the Carter house, where my headquarters had been all day. This position was also in the centre of my command. Ruger's division stretched near half a mile to the west, and Reilly's equally far to the east. Opdycke's brigade was the only organized body of the Fourth Corps infantry now in this space, and it was crowded together so that it and Strickland's brigade only occupied the space of one, except that Russell's regiment, as has been stated, was on the left of the turnpike.

The duration of time in battle is judged so differently by different people that one may well hesitate to guess at it; but it is enough to say that we were in the lull which followed the repulse of the enemy's first assault, the first of a considerable series of persistent attacks which continued on into the night. My own opinion has been that about fifteen minutes had elapsed since the beginning of Opdycke's charge. Others, with equal opportunity of judging, thought the time much shorter than this. Stanley was entirely alone, for he had outridden his attendants, or they had stopped to rally the crowd of fugitives in the streets of the village. Had any of them been present, there would have been no occasion to dismount my staff officer to furnish a horse for the retiring general. These things show how brief must have been the time occupied in the rush and the struggle to retake the works. I saw Stanley ride away toward the village, and I saw him no more till next day. He returned my horse to me then at Nashville.[1]

[1] The importance of these details of time and place will appear later, when I shall have to discuss controversies that have arisen over them. See chap. xxi., *post*. I also postpone the discussion of the numbers of disorganized troops from Wagner's division who rallied at the works and fought with Strickland, Opdycke, and Reilly. One of my staff officers recollects seeing a mounted man meet General Stanley on his way toward the village, and go back with him: whether such person was an officer or an orderly he cannot tell.

CHAPTER VII

THE FIRST FIGHT AT THE CENTRE

Hood's Advance retarded by the Outpost — His Right Wing farthest forward — My Staff at the Carter House — Wagner also there — Messages from the Outpost — Wagner's Replies — Marshall's Guns come in — Disorganized Retreat of the Outpost — Wagner's Efforts to rally — Swept along to the Town — Cannon in the Enemy's Hands — But soon retaken — Fight over the Batteries — Reilly's Second Line charges — Fight at the Cotton-Gin — Destruction of Confederates — Heroism of their Officers -- Reilly's Report — Opdycke's Formation for the Charge — Position of his Regiments — Of Strickland's — Focus of the Fight — Two Lines on Carter Hill — Turn of the Tide.

I HAVE thus far followed the progress of the battle from my personal point of view, telling what came under my own eye, and have necessarily omitted much that had happened at the centre. Let us go back and pick up the thread of events that took place there whilst I was absent on my ride to the left of the line.

The bold skirmishing of Colonel Lane's brigade at Privet Knob, assisted by the good practice of Lieutenant Mitchell's section of artillery, had, from the first deployment of the enemy, produced a visible effect; and when the two brigades of Conrad and Lane took the position where they made their fight, the delay of Hood's advance in the centre became very noticeable. The lines of the two brigades making a wedge with the apex forward, the Confederates showed a corresponding hollow in their formation, so that when the final break came the enemy was nearer my works than the men of the two brigades.[1] Thus it happened that Hood's lines, especially on our left, reached our breastworks almost as soon as the two brigades in front were fully engaged.

Conrad says that the enemy were actually repulsed for a little while, and "fell back under the crest of a small hill" in his front, but quickly reformed and advanced again.[2] It was not till, on this renewed advance, Cleburne's troops were actually over the barricade, and fighting with clubbed muskets had occurred, that the rush to the rear began.[3]

Three or four of my staff had remained at the Carter house, General Wagner had ridden out to Conrad's brigade and returned, and was now resting, dismounted, at the opening in the works where the turnpike passed through the line. My division engineer officer, Captain Scofield, was there, directing some strengthening finish of the breastworks, and tells the story.[4]

"They remarked that the musketry firing was becoming more rapid, also that from the two guns

[1] Twining Map, *ante*, p. 45.

[2] O. R., xlv. part i. p. 271. He and Lane each report that he fell back only after finding the other brigade was retreating. *Id.*, pp. 256 and 271. Such conflict between reports is common in similar circumstances. Their retreat needed no such excuse.

[3] *Ibid.* Conrad made a separate report to the Adjutant General of Missouri for his own regiment, the 15th Missouri. The two reports supplement each other. See Report of the Adjutant General of Missouri for 1864, p. 92.

[4] Captain L. T. Scofield of the 103d Ohio, detailed as topographer, etc. He is a very prominent architect of the city of Cleveland, and a leading citizen there. The extract is from his paper read before the Ohio Commandery of the Loyal Legion, and published in their second volume of "Sketches of War History," pp. 133, 134. The paper is entitled, "The Retreat from Pulaski to Nashville."

in front.¹ By and by a staff officer rode fast from one of the brigades, and reported to Wagner, excitedly, 'The enemy are forming in heavy columns: we can see them distinctly in the open timber, and all along our front.' Wagner said, firmly, 'Stand there and fight them,' and then, turning to the engineer officer, said, 'And that stubbed, curly-headed Dutchman *will* fight them too,' meaning one of his brigade commanders. 'But, General,' the officer (Captain Scofield) said, 'the orders are not to stand, except against cavalry and skirmishers, but to fall back behind the main line if a general engagement is threatened.' In a short time another officer rode in from the right in great haste, and told him the rebels were advancing in heavy force. He received the same order. The officer added, 'But Hood's entire army is coming.' Then Wagner struck the ground with his stick, and said, 'Never mind, fight them.'"

Captain Scofield next describes vividly the appearance of the unfortunate outpost and the pursuing enemy as they surged into sight, with the warning that "Hell had broke loose," which was given by Mitchell's artillerymen as the section came in. Then followed the disordered infantry. "Through the gap, at last, and over the works they came, with Cleburne and Brown hot after them. Wagner by this time was on his horse, riding backward and facing the disorganized brigades, trying as hard as ever man did to rally them. With terrible oaths he called them cowards, and shook his broken stick at them; but back they went to the town, and nothing could stop them."²

[1] The section of Marshall's battery that was with Conrad and Lane.
[2] In a private letter (November 30, 1889), Captain Scofield gave me

Wagner's objurgations in trying to rally his troops must not be taken for his real judgment of their soldierly character. They only proved the heat of his wrath over the visible results of his own mistake. It is true that the regiments had within a few days received large numbers of fresh recruits, and this weakened their cohesive power; but they had each a large nucleus of veterans, who had proved their admirable quality on every field where the Army of the Cumberland had gathered glory, from Shiloh to Atlanta. Even in their false position in front of our main lines, they had stood their ground till Cheatham's corps had nearly enveloped them. If any fault were to be found, it would rather be for an excess of stubborn courage when they knew that "Some one had blundered." Once broken, and military organization lost, a brigade becomes a mere crowd. A rout is then pretty sure to become panicky, and the dictate of sound judgment is that their officers should reform them at some place in the rear, out of the fire, where their organization can be restored; for until then they cannot be handled for any military purpose. The cover for them in reorganizing must be found in reserves that are in unbroken array and can be manœuvred by the tactical words of command. This was the task of

other details of this scene, as follows: "General Wagner was on his horse directly in front of the Carter house, and was making superhuman efforts to check his men. His horse was backing against his will, crowded to the rear by the surging mass of his own soldiers, who were pressing against him in great numbers. . . . While the great mass of the men took to the pike after getting over the earthworks, a great many were crowding along in the open ground east of the pike. I saw a chance there to help a little in checking the retreat, and left Wagner's side to do so. I did not see him again, only as he drifted out of sight with his men toward the town. . . . I know that General Wagner did not return to our lines during daylight."

Opdycke's brigade, lying in column on the right of the road, whilst the broken mass rushed by, and then deploying and charging with a front of steel. It was the task also of Reilly's and Strickland's reserves, restoring the line near the cotton-gin, and making the second line behind the Carter house the rallying place to repulse the enemy. Numbers of brave men in the broken ranks would obey the impulse to turn when they reached the breastworks, and have a hand in the repulse of the enemy, but it would be the plain duty of their officers to collect them for reorganization the moment the lines were held by the organized reserves. This was practically what occurred in the present instance. Wagner, with his principal subordinates, stopped the retreating mass in the town, and, establishing their place of reorganization near the river, where there was room for the purpose, gradually brought into shape the disordered regiments, sending officers to collect and bring to the colors their scattered men.[1]

The statement of Captain Scofield in regard to the orders sent back by Wagner through the messengers that came from his brigades is so completely in accord with the official reports of Conrad and Lane that further corroboration is hardly necessary. One of these messages was delivered to Wagner in the presence of Captain Theodore Cox, my Adjutant General, and he, with Lieutenant D. C. Bradley, one of my aids, also remonstrated with Wagner for sending back orders to fight Hood's army advancing in force. Wagner's excited persistence in his order was his only reply.[2] The messenger himself has

[1] The details of the evidence establishing this will be given in chap. xix., *post*.

[2] Captain Cox's statement was prepared when I was collecting ma-

written his account of what occurred. Conrad and his brigade had the "understanding that when the enemy appeared in heavy force they were to retire inside of the main line." When Hood's army advanced, Conrad sent him to report the fact. He found Wagner at the porch of the Carter house, and made his report. "General Wagner said, ' Tell Colonel Conrad that the second division can whip all Hell, and for him to hold his position'; which I reported to Colonel Conrad, and the Colonel said, 'All right, he would try.' "[1] It is only fair toward Wagner, however, to note that at the last moment he seems to have been recalled by the remonstrances of my staff officers to a consciousness that he was committing an error, and tried to modify his order. Conrad reports that, after all the imperative directions to hold his ground, "just as the enemy got within good musket range, a staff officer of the gen-

---

terial for my volumes of campaign history in the "Scribner's Series," already referred to. As it contains matter pertinent to several incidents of the battle, it is given in full in Appendix F. My brother became the Adjutant General of the Twenty-third Corps with rank of Lieutenant Colonel, and was my Chief of Staff. After the close of the war he returned to civil life, and when he died was Secretary and Treasurer of the Little Miami Railroad Company.

[1] The messenger, Mr. T. C. Gregg, who was taken for a staff officer, was in fact the regular orderly at the brigade headquarters, who had served with General Bradley till he was wounded at Spring Hill, and then with Colonel Conrad. He had been with General Sheridan at Stone's River, was twice wounded at Chickamauga, and left on the field. Taken by the enemy, he was sent to Richmond, and after a time was exchanged. He returned to his regiment (51st Illinois) in time to take part in the assault on Kennesaw Mountain and in the rest of the Atlanta campaign. After the war he became a Justice of the Peace, Clerk of the District Court of Calhoun County, Iowa, and finally Mayor of Rockland City,— a career that shows the good stuff there was in the ranks. Hearing of him accidentally, I asked Mr. Gregg for his knowledge of the facts, and he sent me his pithy statement in reply.

eral commanding the division rode up and said that the general ordered that, if the enemy came on me too strong and in such force as to overpower me, I should retire my line to the rear of the main line of works; . . . but as the enemy was so close to me, and as one half of my men were recruits and drafted men, and knowing that if I then retired my lines my men would become very unsteady and confused, and perhaps panic-stricken, I concluded to fight on the line where I then was. So I ordered the men to commence firing."[1]

Two batteries of artillery had been kept in reserve near the Carter place, and, just before the battle opened, Captain Bridges placed his own battery (Illinois Light Artillery) on the right of the turnpike, near the middle of Strickland's brigade line. One section of Battery A, 1st Ohio, was sent to reinforce the 20th Ohio Battery, just west of the Carter house, in the épaulement there, and Lieutenant Scovill of the former was directed to take command of all six guns, as the only officer with the 20th was a junior lieutenant, Mr. Burdick.[2] The other section of Scovill's battery was placed close to the turnpike on the left, to rake the road if the enemy should break through.

When Wagner's two brigades broke away from the ridge in front on which they had been placed, the 1st Kentucky Battery, which was at the immediate left of the opening in our works, was of course masked by the crowd of fugitives, and could not be put in action. Nothing is more helpless than a battery in such circumstances. It occupies a place where infantry might possibly do something if they

[1] O. R., xlv. part i. p. 270.
[2] Bridges's Report, *Id.*, pp. 320, 321.

were there, but in their absence, and unable itself to fire, it only adds to the width of the open door through which our broken brigades and the enemy with them could rush pell-mell. The limbers and caissons retreated, but the guns were, as we have seen, in the enemy's hands, until the counter charge of Reilly's and Opdycke's men regained them. For a little while the guns were worked by volunteers from Colonel Russell's regiment (that on Opdycke's left),[1] but the gunners were soon collected again, the caissons brought up, and the battery was steadily worked till dark, firing over a hundred rounds.[2] The enemy hung on stubbornly for a while to the outside of the breastwork, but then surrendered and came over the works as prisoners.

Lieut. Colonel Hayes, commanding the 100th Ohio, which had been next to the guns, tells us in his report, that the enemy swarmed in on the road, and where the battery was, taking his line in flank, as well as following Wagner's men, who trampled over it.[3] He himself shouted to the fugitives to rally at the rear, and his own men, thinking the order was meant for them, broke away; but he says that his color sergeant (Baldwin) led the rally, and placed the colors on the works, though he did it at the cost of his life. The acting Major (Captain Hunt) was also killed fighting at the parapet.

Lieut. Colonel White of the 16th Kentucky led his men into the space occupied by the battery, and, Colonel Russell following him close, the line was made at least four deep, and their fire destroyed the Confederates who did not seek a momentary protec-

---
[1] O. R., xlv. part i. p. 246.
[2] Thomasson's Report, *Id.*, p. 326.
[3] *Id.*, p. 419.

tion in the outside ditch. In like manner the 8th Tennessee and the 12th Kentucky under Lieut. Colonel Rousseau doubled on the line of the 104th Ohio, the right wing of which had also been borne away, and the line of Reilly's brigade was made too solid to be shaken again, even for a moment. Rousseau says he saw the men leaving the works at the angle near the cotton-gin when he ordered his men forward, and it is impossible to tell exactly where the line of the break began.[1] The rally was so prompt, and the rush forward was so determined, that it was but a moment till every Confederate inside the works was dead or a prisoner.

The section of the 6th Ohio Battery, which was in the salient angle at the cotton-gin, was never for a moment in the enemy's hands. Lieutenant Baldwin, who commanded it, gives a graphic picture in his report of the hand to hand conflict there.[2] The enemy tried hard, he says, to force a passage at the right embrasure of the battery. They several times got into the embrasure, pushing their guns through, and firing upon the cannoneers. "They were so unpleasantly close that we had to resort to the use of sponge-staves, axes, and picks, to drive them back."[3]

The left of the second line of Reilly's brigade had been advanced before there was any break in front, and

[1] O. R., xlv. part i. pp. 415, 416.
[2] *Id.*, p. 334.
[3] *Ibid.* Mr. Baldwin adds: "No less than thirteen stand of colors were captured on the battery front by General Reilly's brigade, one of which by right should, as a trophy of that bloody engagement, be in the possession of the battery." I sympathize with the gallant artilleryman, but no commands kept the flags as trophies. They were sent to Washington. The battery fired 550 rounds of ammunition in the battle.

the line of Rousseau's 12th Kentucky was oblique, ready to dash upon the flank of any force coming over the works. It charged in this way, and mingled in the hand to hand fight at the parapet, and around the battery. One of its officers was seen doing the work of a cannoneer in his enthusiasm, with foot and hand on the spokes of the gun carriage, helping the swift handling of it.[1] A Confederate soldier thought to hinder the loading of the gun by thrusting a fence rail into it, and only added to the destruction caused by the discharge. Two of the companies of the 12th Kentucky were armed with revolving rifles, and this was just the pinch when this multiplication of efficiency in the weapon told with decisive effect. Human courage could not endure the fire. An officer looking out under the smoke says the Confederates on the brink of the ditch were as thick as sheep in a pen, tumbling about in a confused mass. Captain Brown of the same regiment seized the flag and leaped to the top of the works, waving it, and shouting. Strange to tell, he leaped down again unhurt. Ammunition ran short. The Adjutant rushed a detachment of men down a ravine to the left, where ammunition wagons were, and in a few minutes they came back, each with a box of cartridges on his shoulder. As these were thrown down, the soldiers fell upon them with axes, and split the covers off to save time in their distribution. The same officer, with others, carried hatsful of the cartridges along the line. What was happening at this point was going on everywhere; the enthusiasm and desperate courage were the same.[2]

[1] From a written statement of Captain Thomas Speed, giving details not included in his printed article before referred to.
[2] *Ibid.*

Colonel Sterl of the 104th Ohio reports with clearness the condition of the front line, of which it formed the centre and left in Reilly's brigade. The salient in the breastworks near the cotton-gin gave an oblique fire in both directions, and his three companies on the left opened first on the enemy advancing against Casement and Stiles, as Loring's division was a little ahead of Walthall's, for reasons that will presently appear.[1] Almost immediately, however, the retreating brigades of Wagner's division poured over the works from the salient westward. These "had scarcely crossed our works until the ditches in front were filled with rebels scrambling to get over. . . . The confusion and hurry of the crossing of this advance line, their officers crying to them to ' get to the rear and reform,' came near throwing our line into confusion, and the three right companies, borne back by them, and in doubt as to the commands, fell back a few paces, but, in almost a moment afterward, rushed back with fixed bayonets and regained their works."[2] He tells us that our second line rushed up simultaneously with them, and all together "kept up a constant and destructive stream of fire, cutting down by hundreds the rebels who had accumulated and massed in the ditches and immediately in front."

Colonel Rousseau adds another telling incident, which shows the energy and heroism of the Confederate officers. The hostile battle-flags were on our works, and "a number of their men had gained the top and fired down into our ranks. Even bayonets and clubbed muskets were used." When these were shot down or hurled back into the ditch, their officers were heard exhorting the men to stick to the

[1] See chap. xi., *post*.   [2] Sterl's Report, O. R., xlv. part i. p. 421.

outside of the parapet, shouting that "they had them, if they knew it."[1]

The briefer description by General Reilly is in close accord with the consistent and clear details given by his subordinates. He speaks also with praise of the conduct of the 175th Ohio, a new regiment, only for that day attached to the brigade, and from which we have no formal report. They came up with the rest in the crisis of the fight, and gallantly shared with the others the glory of restoring and holding the line. They had been on their way across the river in the afternoon, when Lieut. Colonel McCoy, their commander, went in person to General Schofield, and begged permission for his regiment to stay and support the line. Reilly, with Colonel Hayes and Colonel White, heartily recognized the help given also by Colonel Russell with his 44th Illinois of Opdycke's brigade,[2] who came up "right nobly" close to the turnpike, among the guns of the Kentucky battery, which, as we have seen,[3] was recaptured and turned on the enemy. Anticipating the questions which might be mooted, Reilly closed this part of his report with careful explicitness by saying, "that with the exception of the aid rendered by the 175th Ohio Volunteer Infantry and 44th Illinois Volunteer Infantry, as herein stated, the brigade received no assistance during the fight, unless, perhaps, some of the men coming in over the works may have rallied in or behind the lines."[4]

[1] O. R., xlv. part i. p. 416.
[2] Reilly's Report, *Id.*, p. 412.   [3] *Ante*, p. 109.
[4] O. R., xlv. part i. p. 412. General Reilly retained command of his own brigade, whilst also in temporary command of the division. His brigade report is fuller than that for the division, as, after writing the first, circumstances combined to give him no time to do much more than transmit the other brigade reports. *Id.*, p. 410. No brigade

In gallantly leading the 16th Kentucky forward, Colonel White was severely wounded in the face, but he kept on without halting, and remained in command of his regiment till the brunt of the fight was over. He made no mention of his wound in his official report. The casualties in the brigade (not including the 175th Ohio) were two hundred and thirty-two. Six officers and twenty-seven men were killed, seven officers and one hundred and twenty-three men wounded, and one officer and sixty-nine men were missing. An analysis of these casualties affords instructive evidence of the extent of the break under the rush of the retreating brigades of Conrad and Lane. I postpone it, however, till we shall have traced the progress of the fight along the whole line.[1] The order of movement on the part of the Confederate divisions can also be more intelligently considered a little later.

Whilst speaking of the fight near the cotton-gin, I may be permitted to make special mention of one young hero who fell there, a brilliant type of the volunteers who shed their blood for the national cause. Lieutenant James Coughlan of the 24th Kentucky had been one of my aids through the campaigns of the year. Of humble birth, and in the main self-educated, he developed military talents of a high order. His eye was quick, his judgment clear, his courage indomitable, his soldierly bearing

---

commander bore a better reputation than General Reilly. He was an excellent disciplinarian, always with his men, keeping his eye on everything, and leading them admirably in battle. He had been promoted from the colonelcy of the 104th Ohio for distinguished services as brigade commander in the Atlanta campaign. After the war he resumed his profession as a lawyer and afterward a banker, and is one of the best known citizens of eastern Ohio.

[1] See chap. xvi., *post*.

inspiring. His reports were so true and just that he gave his chief just the help needed by one who cannot be everywhere, yet must know what is happening. He had carried my first order to Opdycke to be ready to charge to the front in case of trouble, and on his return dashed into the thickest of the melée with his old comrades of Reilly's brigade, conspicuously cheering them on to regain their breastworks. He fell near the cotton-gin in the moment of victory, and there his fellow staff officers and friends buried him in the darkness of the night, in the intervals of the fierce fighting.

The charge of Opdycke's brigade was not made in a single line, but as the brigade had been lying in reserve on the right of the turnpike in column of regiments, the leading regiment became the centre in the advance, the others taking ground to right and left in echelon on its flanks. This leading regiment was the 88th Illinois (with which the 74th had recently been consolidated), and was commanded by Lieut. Colonel Smith,[1] of whom Opdycke says in his report that he "was conspicuous, even among heroes." In rear of Smith, on the left, were the 73d and 44th Illinois respectively, and on the right the 24th Wisconsin, 125th Ohio, and 36th Illinois.[2]

It is impossible to tell, from the evidence which has been preserved, what was the line of march of the leading regiment, though the place of each, when the line in front was restored, is definitely fixed.[3]

[1] O. R., xlv. part i. p. 241.

[2] *Id.*, p. 251. Opdycke's report places the 24th Wisconsin on the right of the 125th Ohio, but as the official report of the latter explicitly places it between the 24th Wisconsin and the 36th Illinois, I follow this order. In such details the knowledge of the regimental commanders is most immediate.

[3] *Ante*, p. 98.

We have already found Russell, with his 44th Illinois, among the guns immediately on the left of the turnpike. The 73d Illinois reports that it "reached the works now upon the right of the pike[1] just in time to drive the enemy back and save a battery which had been left without any support." Next on the right was Smith's consolidated regiment, and beyond him the 24th Wisconsin. The reports of these regiments speak of charging up to works. Still farther on the right came the 125th Ohio, whose commandant (Captain Bates) reports that "two guns at the right of the regiment that had been deserted by all but a single corporal were quickly brought into action again, new barricades constructed, stragglers forced back to them, and disaster averted."[2] Last on the right flank of the brigade came the 36th Illinois.

The position of the guns recaptured is conclusive proof as to the positions taken up by these regiments. The guns nearest the turnpike on the right were the 20th Ohio Battery, which had been placed "just west of Carter's house" by my own direction, and their épaulement was close on the right of the brick smoke-house.[3] The two guns farther west were a section of Bridges's battery, which he as chief of artillery had, just before the battle opened, "placed near the centre of Strickland's brigade."[4] There was no artillery beyond these guns till we come to the Pennsylvania battery on the Carter's Creek Turnpike, on the extreme right of Ruger's division, and nearly half a mile away.[5]

---

[1] O. R., xlv. part i. p. 248.   [3] *Id.*, p. 351.
[2] *Id.*, p. 251.   [4] *Id.*, pp. 320, 351.
[5] General Ruger, in his report (*Id.*, p. 365) says, that he placed a section of the 20th Ohio near the right of Strickland's brigade. The

The First Fight at the Centre 117

It thus becomes tolerably clear that Opdycke's regiments, going forward in a wedge-shaped formation by echelon, doubled their ranks, and occupied a little more than the left half of Strickland's brigade line, together with the retrenchment on the Columbia Turnpike, the ditch along the east side of the road, and some space immediately on its left. To reach their positions by direct lines from the place where they lay in reserve, Colonel Smith must have led forward just in rear of the Carter house, through the yard, and up to the outbuildings. The other regiments partly overlapping him and each other came in on right and left. The house, of course, prevented an advance in continuous line, but it also, with the outbuildings, had made points of rallying and stubborn defence for some of Strickland's men and some of those retreating from the front. The troops on the turnpike had no obstructions in their way, and could move most rapidly. The break in Reilly's line had been first closed up, and from the salient at the cotton-gin as well as from the breastworks near, an oblique fire swept across the road and down between the first and second lines, making it easier to regain possession of the interior line of works beyond the outbuildings, which had been built by the 44th Missouri, and which Lieut. Colonel Barr declares in his report they never left.[1]

---

artillery reports show that this was modified by Lieut. Colonel Schofield and Captain Bridges, chiefs of artillery. As the 20th Ohio had only one officer with it, it was kept together, and Lieutenant Scovill of Battery A, 1st Ohio, was sent to command it. A section of Captain Bridges's own battery was placed near the centre of Strickland's brigade just before the action opened, and about the time of Bate's attack was reinforced by the other section, and with Ziegler's Pennsylvania battery crossed fire on the advancing enemy. Bridges's Report, *Id.*, pp. 320, 321; Scovill's Report, *Id.*, p. 330; Sergeant Horn's Report, *Id.*, p. 336.

[1] Colonel Bradshaw was very severely wounded in the battle, and

In Strickland's brigade, the 50th Ohio, which was the left of his front line, was driven back on the Carter buildings, and rallied with Opdycke's men as these advanced. The 72d Illinois, which was the right of the same line, held fast to the breastworks on the extreme right; but the left wing of the regiment was swept back, and rallied upon the 183d Ohio, the new regiment which was its support, and continued the second line of breastworks built by the 44th Missouri (as part of Opdycke's men were also doing) by throwing together fence rails and any other material out of which a barricade could be hastily made.[1] The right wing of the regiment was ordered back by the Colonel to join the left in second line,[2] and to them also rallied larger numbers of men from Wagner's broken brigades than were found elsewhere in the line.[3]

All the circumstances show that the gap west of the Carter house was longest open, and that bodies of the enemy got farthest within our lines there. General Cheatham says that on the next day they found some of the Confederate dead fifty yards within the breastwork at this point.[4] As they advanced, however, they had Opdycke's charging regi-

---

the report was made by Lieut. Colonel Barr. His fullest statement is in his report to the Adjutant General of Missouri (1865, p. 276). My references will be to that when I speak of his report. See also sketch map, p. 43, *ante*.

[1] O. R., xlv. part i. p. 354.
[2] Sexton's Report, *Id.*, p. 393.
[3] Statement in writing by Captain Sexton, who commanded the regiment after Lieut. Colonel Stockton and Major James were wounded. I shall have occasion to refer to this again in estimating the number of Conrad's and Lane's men who rallied at the line.
[4] Private letter to Colonel Opdycke, of which I have a copy. Also statement by General Gordon in his memorial address, hereafter mentioned. See chap. xi., *post*.

ments on one side with the 44th Missouri, and on the other the 72d Illinois and 183d Ohio, crossing their fire upon them, and as they hesitated these troops charged upon them and they fell back in confusion to the outer side of the main line of breastworks.

The retreat of Strickland's men from the main line had left the flank of Moore's brigade exposed, but Captain Dowling, the brigade inspector, gallantly led two companies to the flank of the brigade and placed them in position to sweep with their fire the open space between the lines. The enemy was thus driven from close proximity to Moore's left, and his line never budged from its position.[1]

That the focus of the fight was around the position of the 44th Missouri, just in rear of the Carter house, is shown by the list of casualties in that regiment, which had more men killed than all the other regiments of the brigade.[2] The distance between the two lines at the turnpike was some sixty-five yards, growing less toward the west,[3] and across this space the fire leaped from the muzzles of the rifles and of the cannon, now hotly in action again. The men in and about the Carter buildings were better covered than the rest, though, from the nature of the case, they were more broken in their order. They fired from the windows of the buildings and from every opening or interspace that could be used as a loop-hole. They clustered at the corners and between the out-buildings, and fired obliquely from this cover.

---

[1] See also chap. ix., *post*.

[2] Its losses were 30 killed, 35 wounded, and 86 missing, total 151. The losses of the whole brigade were 47 killed, 150 wounded, and 281 missing, total 478. O. R., xlv. part i. pp. 368, 369.

[3] *Id.*, p. 354.

From both Reilly's line and Moore's, oblique fire also crossed the deadly field, and the balance of fortune turned heavily against the Confederates. It was not now the case that "they had the works if they only knew it," as their officers had shouted in front of Reilly: the truth rather was that they had lost the battle if they only knew it, and it would have been wiser to have drawn off the shattered battalions. But they did not know it. The smoky battle-cloud so hid the field that Hood and his corps commanders long thought our lines were irretrievably broken, and pushed forward their troops to slaughter.

With the new establishment of a defensible line at the centre, the first step toward complete success had been well taken, and though the lull which followed was to be short, it will give us time to trace the current of the fight on either flank of our position across the whole front from river to river.

CHAPTER VIII

THE FIGHT OF OUR LEFT WING

Advance of Stewart's Corps — The narrowing Field — Thorny Hedges — Changes in the Array — Walthall's Report — Loring's Division strikes Stiles — Fighting in the Railway Cut — Batteries at Close Range — Loring Repulsed — The Attack on Casement — General Adams's Death — Our Line successfully held.

I HAVE already described the manner in which the centre of Hood's line of battle was momentarily checked by the fire of Conrad's and Lane's brigades at their outpost, and how the enemy's wings swung forward.[1] It thus happened that the fierce attack on our works fell first on Stiles's and Casement's brigades in Reilly's division; for Wagner's two hapless brigades were nearly half a mile in front, and their useless struggle gave time for Stewart's corps to outmarch Cheatham's. We have followed also the fate of Wagner's men as the natural key to the events of the battle till the imminent peril at our centre was passed.

It would have been more in accord with the actual order of events if we had begun our examination of the history of the battle at our extreme left, in Stiles's brigade; for it was there that Hood's infantry first struck our main line. But the connection of cause and effect is, in this case, more important than the sequence of the hours. The position assumed by

---

[1] *Ante*, p. 103.

Wagner as an outpost gave its peculiar character to the battle, and its detailed progress on other parts of the field could only be properly understood when we had seen the results of the Confederate attack upon the two brigades, whose rout seemed to open the way into the heart of our position.

We may now go back and trace the progress of the attack upon our two wings in succession: first upon Stiles's and Casement's brigades in Reilly's division, and then upon Moore's brigade of Ruger's division, lapping upon Kimball's division where Chalmers's Confederate cavalry extended the enemy's efforts to our extreme right flank. On these parts of the line was fairly tested the ability of the Southern army to carry such fieldworks as ours by an assault over open ground, without the advantage of following close behind a broken and retreating body of our own men. The result was not encouraging to such attacks.

The formation of the Confederate corps under Stewart was, as we have seen,[1] a line of battle of three divisions, each division formed of two brigades in front with the third brigade marching in reserve. The exception was his left division under French, which was formed with one brigade in front and one in second line, the third brigade (Ector's) being absent, guarding the pontoon train.[2] The right of the corps was near the Harpeth River, which here ran northwestward till it reached the flank of our works. The width of the field of battle contracted rapidly as Stewart advanced, and it had, besides, several obstacles which were not visible at a distance, or did not seem formidable till one was upon them. Of these the chief was a field which had been

[1] *Ante*, p. 87.   [2] O. R., xlv. part i. p. 708.

surrounded by the quickset hedge of Osage orange, of which mention has already been made.[1]

Walthall's division was the central one, and he alone made any full report of the engagement, so far as the official records show.[2] He describes with clearness how the thicket and hedge on the farther side of the great field through which he advanced made it necessary for his left brigade (Reynolds's) to get around it by taking ground to the left, and on passing it Reynolds was not able to get back to his place in line. Walthall therefore brought up Shelley's brigade, which was in reserve, and put it in the gap. A short halt was made to reform, and then the command to charge was given. As the Confederates of this wing were guiding by the right flank and along the river bank, both Loring and Walthall had to take ground to their left as they advanced, and this brought Walthall in front of Cleburne, delayed as the latter was by the resistance of Wagner's men. Walthall's left seems to have run over part of the outpost line held by Conrad, or to have gone so near to it that he speaks as if he had had a part in routing it.[3]

It was Loring's division, therefore, with the right of Walthall's, that fell upon Stiles and Casement. As soon as the enemy's line came up into view upon the open plain in front, it was opened upon by the two batteries on the knoll at our left, as well as by one which had been placed in Fort Granger.[4] Stiles's

---

[1] *Ante*, p. 53.   [2] O. R., xlv. part i. p. 720.   [3] *Ibid.*

[4] This was Battery D, 1st Ohio Light Artillery, composed of 3-inch rifled ordnance guns. It was admirably adapted for this long range work, and was under the eye of Captain Cockerill, who was also Chief of Artillery of my division. It expended 160 rounds of shell and three rounds of case-shot in assisting to repel this attack of Loring upon Stiles. See O. R., xlv. part i. p. 432.

left regiment (120th Indiana) was sharply recurved along the railway cut, which was quite deep, and the regiment was exposed to an enfilading fire of both cannon and small arms, but held its place firmly. The centre of the brigade was covered by the thorny hedge parallel to the front, and the enemy, charging up to this, was dismayed and exasperated at finding it impassable, whilst the small tough trunks of the trees, thinned out as they had been, did not protect them from the destructive fire of Stiles's line.[1] On the right of the brigade the first assault of Loring's men met a bloody repulse. His second line soon came on and reached the works in front of the 128th Indiana. They planted their colors on the parapet, and some climbed over the works. The color bearer was shot down, and the men who had got over were made prisoners. The obstinacy of the Confederates was met by an equal pertinacity of courage, and the assault was repulsed.[2] Then the rebel officers made strenuous efforts to press their men to their right, and to get between the river and our left. "They were heard to exclaim, 'Soldiers, as you love your country, press to the right! press to the right!' But the terrific and well directed fire of the battery of regulars (M, 4th U. S. A.) and the musketry fire of the brigade kept them in check."[3]

[1] The hedges perpendicular to the front had been cut away to extend an abattis toward the turnpike, and the farther one, across the great field, was that which had disarranged the advance of Walthall.

[2] Stiles's Report, O. R., xlv. part i. p. 430.

[3] From the statement of Colonel Henderson which I have more fully mentioned in chap. x., *post*. To explain Colonel Henderson's position it should be understood that he had been seriously ill, but had remained with the brigade, though Colonel Stiles of the 63d Indiana was temporarily assigned to the command. This condition of things had lasted for several days; yet at Franklin he could not stay away from the front, and, weak as he was, was on the line. Technically, Colonel Stiles

The regular battery (Lieutenant Samuel Canby in command) swept the field in front of the brigade till the enemy were close to the works, and then its right section was placed on the left of the left section, where it could enfilade the railroad cut in front of Colonel Prather's regiment (120th Indiana) as well as the river bank. Colonel Henderson says, "The battery of regulars behaved so gallantly that my admiration was excited." It did most effective service.[1] But Canby's battery was not alone on the left flank. When Mitchell's section of Battery G, 1st Ohio, came within our works at the centre, it joined the rest of the battery, and Captain Marshall was sent to reinforce the left, near the railroad, just as the attack on Stiles's brigade was fully developed. He carried still further the extension of artillery to the left, which Canby had made by the transfer of his right section, and opened with canister, enfilading the railroad in front of the original position of the regular battery. He tells us that the enemy were there in strong force, and coming up to our works on their hands and knees.[2] The rifled guns in Fort Granger, across the river, were also in rapid action, and the sharp, continuous, grinding rattle of the musketry combined with the cannon roar to tell why it was that Scott's and Featherston's brigades of Loring's division had so terrible a list of casual-

remained in command, but Henderson gave assistance in everything as he was able, and the cordial good understanding between the two officers gave the brigade the benefit of the skill and courage of both.

[1] I take the more pleasure in quoting Colonel Henderson's praise because Lieutenant Canby's official report is a very brief and modest statement of the battery's part in the action. See O. R., xlv. part i. p. 338.

[2] O. R., xlv. part i. p. 331.

ties.[1] The fight here was bitter and stubborn. It is difficult to determine how long it lasted. The enemy's reserves renewed the action at least twice, but at last they gave it up, and those who could get away straggled back to the hollows in front to reform. Numbers surrendered in the ditch, and came over the works as prisoners. Others lay flat along the ground in front of the hedge, and waited for darkness to cover their retreat. At the end of an hour from the opening of the fight, Stiles's front was clear of aggressive bodies of the enemy, and in the dusk of the early evening a skirmish line was advanced, and kept out till the line was withdrawn at midnight. During the advance of Loring's division, Guibor's Confederate battery had hotly engaged our artillery at the extreme left, and had made the position of the 120th Indiana, formed as it was *en potence*, facing the east, a most trying one. When, however, the contest of the infantry became close, the enemy's guns were withdrawn from the range of Fort Granger.

The fighting on the line of Casement's brigade was similar to that upon Stiles's front, but as the line itself was straighter, and had no recurved flanks, the contest was more even, one regiment being no more exposed than another. So also no part of Casement's line had any advantage, such as the tough standing hedge gave to Stiles's centre. The whole brigade front was covered with the light abattis made from the hedge material, but this was not pinned down, and was more a show of obstruction to the enemy than a reality. The flanking company of the 65th Indiana (which was the extreme right of

---

[1] Loring officially reported an aggregate of 876 casualties in his division. O. R., xlv. part i. p. 715.

the brigade) was armed with repeating rifles.[1] The head-logs on the parapet were perhaps more continuous than in some other brigades in the line, but the chief advantage of Casement's position was that it had clear range in front, and was just beyond the disturbing influence of the rush of routed men from the outpost.

No officer of the division had a stronger personal influence on his men than Casement. At once impetuous and clear-headed, he was everywhere present, his ringing voice heard above the din. His men were well in hand, and opened fire as soon as the enemy came within reach of their rifle balls. It was probably the left wing of Featherston's brigade which first struck this part of our line. The struggle was sharp, but the attack was repulsed. The Confederates reformed and tried it again, but again and still again they were driven back, leaving the ground covered with their dead and wounded. Casement reported that "the firing was kept up with great vigor until dark, during which time the enemy made several distinct charges, but were repulsed each time with terrible slaughter."[2]

In one of the lulls between these attacks, when the smoke was so thick that one could see a very little way in front, the officers of the line seized the opportunity to look over the parapet, the better to learn the situation. The enemy's line was dimly seen, and a mounted officer in front of them, upon a fractious horse, either forming them for a fresh

---

[1] It was only toward the close of the war that repeating and breech-loading arms were issued, and at first in so small numbers that only a company here and there was supplied with them. In such cases the company was the ranking company of the regiment, which stood on its right.

[2] O. R., xlv. part i. p. 425.

assault or rallying them after the last repulse. Several shots were fired, and horse and rider both fell. Presently the horse struggled to his feet, and dashed wildly forward, straight for the breastwork, leaped upon it, and fell dead astride of it. The officer, shot through the thighs, tried to crawl away. Casement, who was at the moment at the extreme right of his brigade, shouted to him to come in and save his life; but he kept on, probably not hearing the well meant advice in the horrid noise, and flying shots pierced him again. He lay helpless till, the repulse of his command being complete and darkness coming on, Casement ordered forward a skirmish line to cover his front. The wounded officer was General John Adams, whose brigade was the left of Loring's division. He was brought in still living, and immediate surgical care was given him; but his condition was past help, and he soon died.[1]

The incident has historical importance, as fixing Adams's place nearly opposite the right of Casement's brigade, and showing that he probably came up as a reserve behind Walthall's right. Four Confederate brigades thus attacked two of ours, and our line was shortened by having two regiments in reserve. Neither in Stiles's brigade nor in Casement's was there any need to call upon the reserves. Those of the Confederates were all brought into action, and

---

[1] O. R., xlv. part i. p. 353. In my official report I followed the account first circulated, which was that General Adams mounted the parapet in a charge, and fell there. I was set right in this matter by written statements of General Casement and of Captain Hornbrook of the 65th Indiana, made independently and in substantial accord. Adams's saddle with the bullet holes in the flaps was long a relic in General Casement's possession. His watch and other valuables were sent to his family under a flag of truce, in our North Carolina campaign in the next spring.

the short line which they attacked made their flanks lap upon each other. In this way was produced the effect of an attack in columns several lines deep, which was the description of the attack universally given by our officers and men. The front line, when repulsed, rallied on its supports, and the efforts to storm the intrenchments were thus renewed repeatedly as long as daylight lasted. Casement's losses were slight, and he touches the true cause of this when he says in his report, "Not a man left the works unless ordered to do so, which accounts for the small loss."[1] The experience of the brigade goes far to prove that, even with muzzle-loading rifles such as our troops were then armed with, the protection of fairly good earthworks is sufficient to enable the line formed in two ranks to cripple so seriously an enemy twice as strong, that his aggressive impetus is lost, and his attack fails when made over open ground. And this was done with a loss to the defenders scarcely worth mentioning.[2]

[1] O. R., xlv. part i. p. 425.

[2] The same lesson was taught at Cold Harbor in Virginia, where the sides were reversed. Casement's casualties in the brigade were only nineteen. O. R., xlv. part i. p. 411. Those of Loring's division have been mentioned above, p. 126.

CHAPTER IX

THE FIGHT OF OUR RIGHT WING

Cheatham's Corps — Convergent Attack of Cleburne and Brown — Line of Lane's Retreat — Moore's Brigade Front unmasked — Our Artillery Cross-fire — Advance of Bate's Division — Battery at the Bostick Place — Close Quarters at Moore's Centre — Help from Kimball — Chalmers's Cavalry attack Kimball — Infantry attack his Left — Cavalry his Centre and Right — Confederates repulsed.

THE right of Ruger's division (Moore's brigade) was not engaged quite so early as Strickland's brigade, and the circumstances were somewhat different. The wedge shape of Conrad and Lane's lines caused the attack of the Confederate divisions of Cleburne and Brown to become convergent, for their charge came straight at the front of the two brigades resisting them. When these broke, the lines both of retreat and pursuit all led toward the turnpike and the opening in our breastworks there. In all the enemy's regiments which attacked Lane, therefore, the left shoulder was thrown forward, and that flank was turned obliquely to Moore's line. The fire opened as soon as the Confederates were unmasked by our retreating men, and it told with terrible effect. No part of the fugitives made for the breastworks here, for, as we have seen,[1] their path was bounded by the centre of the 72d Illinois in Strickland's right wing. Moore's men proved, as did Casement's and

[1] *Ante*, p. 118.

Stiles's on the east of the turnpike, that there was no doubt of their ability to hold the works where there was any scope for their fire, and where the direct impact of the flying crowd was not added to the advantage the enemy had in coming to the very ditch under such a cover.

When the front line in Strickland's front gave way, the left of Moore's brigade was seriously compromised, but the courage and determination of officers and men were equal to the task. Colonel Sherwood, commanding the left regiment (111th Ohio), had his men fix bayonets and prepare for a hand to hand fight on the parapet.[1] Captain Dowling, the Inspector of the brigade, rallied some of the broken troops by heroic efforts, and led them into position as a flanking force to cover the left of Moore's line.[2] A little later, when the enemy was making renewed efforts to break through, part of the 101st Ohio, from Kimball's division, was sent by my order to reinforce Ruger's line. With these, Captain Dowling established a short line nearly at right angles to the main one, where they swept with their fire the space in front of the new line of Strickland's men, and drove the enemy from the works in the immediate vicinity of Sherwood's flank. Dowling was wounded in the struggle, but his work was well done, and the line thus reinforced remained unshaken.

Ziegler's Pennsylvania battery on the Carter's Creek Turnpike was ordered to fire obliquely to the left, where the enemy were seen to be forming upon the elevation on the farther side of the hollow already described, which meandered in front, and

---

[1] Sherwood's Report, O. R., xlv. part i. p. 387.
[2] Moore's Report, *Id.*, p. 380.

ran diagonally through our lines near the extreme right. Bridges's Illinois battery, which was on the slope near the centre of Strickland's brigade, crossed its fire with Ziegler's, and the infantry also carefully directing their rifles against the same height, the Confederate line, as General Ruger says, visibly wasted and disappeared under the concentrated fire.[1]

The first attack upon the front of Ruger's division had been made by Brown's division of Cheatham's corps, which had marched forward, guiding its left upon the Columbia Turnpike. But for the reason already stated, the Confederate brigades had changed their direction more toward our left. Gordon had partly crossed the turnpike, and his brigade was astride of it when he struck our works. Gist had extended the line so as to involve Moore's brigade on our side in the attack.[2] Strahl and Carter respectively supported Gordon and Gist. These seem to be the troops who, after their repulse, were reforming beyond the hollow where Lane's brigade had stood, and who were withered in the fearful fire.[3]

Bate's division of Cheatham's corps had been deployed on Brown's right and rear. It was formed with two brigades in front line, Jackson's on the right, and Smith's on the left, with Bullock's in second line.[4] Presstman's battery was also with him. Bate had a longer road to travel than Brown, and this made his attack somewhat later.[5] His advance was guided by the house of Mrs. Bostick, near the

[1] Ruger's Report, O. R., xlv. part i. p. 365.
[2] Capers's Report, *Id.*, p. 736.
[3] The evidence for this will be given when I analyze more fully the lines of advance from the Confederate standpoint, *post*, chap. xi.
[4] The permanent commanders of the two brigades last named were Generals Tyler and Finley, but they were both absent.
[5] Bate's Report, *Id.*, p. 743.

Carter's Creek Turnpike, and when he reached it he had one brigade on each side of it, his left being close to the road. Presstman's battery went into action on the knoll by the house, whilst Bate, finding a stout line of skirmishers on his left flank (from Kimball's division as well as Ruger's), moved his reserve brigade to the left so that it was astride of the Carter's Creek road with two regiments on the west of it. He had found his left under a furious fire from Ziegler's battery and the right of Ruger's line, and this further deployment was to give the needed extension to his own front. He expected Chalmers's cavalry division to have joined his left, but Chalmers was much farther around, in front of Kimball, and was advancing nearly at right angles to Bate's line of march. The latter therefore thought himself entirely unsupported.[1]

The tendency to move by convergent lines is seen in Bate's as well as in the other Confederate divisions. His right swept over the "half-moon" barricades that Lane had held, and came up to our works, where Gist's and Carter's brigades of Brown's division had preceded him. His centre came full upon Moore's brigade, and his left, beyond the Carter's Creek Turnpike, lapped upon Grose's brigade of Kimball's division.

The struggle was a severe one, and in several places along Moore's front the enemy came to close quarters over the breastworks. The centre of the brigade was occupied by the small detachment of two companies Ruger had taken from the 183d Ohio to fill a gap in his attenuated line. These men were raw recruits, and in the pressure gave way. Moore took two companies from the 80th Indiana (on his

[1] Bate's Report, O. R., xlv. part i. p. 743.

extreme right), and hurried them to the centre, where they made good the momentary gap. At Ruger's request, I also ordered Kimball to send a regiment to reinforce him, as has been already stated. A full regiment was not at the moment available, and Lieut. Colonel McDanald was sent with part of the 101st Ohio.[1] These were put in at the weak place on the left of Moore's brigade, and did brave and efficient service.

At Moore's centre, where the enemy had gained some brief advantage, they made desperate efforts. They made a bold dash for the flag of the 107th Indiana, which was on the works a little to the left of the centre, but those who grasped it were shot down before they could accomplish their purpose. Battle-flags were planted on the parapet, but the 23d Michigan turned its fire obliquely to the left, and shot down two color bearers and cleared the works of the assailants. The space was immediately filled by the two companies of the 80th Indiana hastening by orders from the right, and the line was no more in danger.[2] The sharp fighting was, however, confined to Moore's centre and left. His right, which was recurved, was not at close quarters with the enemy; for Bate's left brigade kept its alignment with the centre, and did not reach our works at the Carter's Creek Turnpike by a hundred and fifty yards. The casualty returns support this view. The two regiments on the right (80th Indiana and 118th Ohio) report nine men wounded only. The total loss of the brigade was one hundred and ten,

[1] General Ruger says five companies. Colonel Kirby, commanding Kimball's brigade, says eight. O. R., xlv. part i. pp. 367 and 184.

[2] Moore's Report leaves it doubtful what troops supported the left of his line and the centre respectively, but the reports of the 23d Michigan and 80th Indiana seem to make it certain. *Id.*, pp. 383, 386.

of which only fifteen were missing. The heaviest loss fell naturally upon the 111th Ohio, which was exposed by the break in Strickland's brigade. This regiment reported forty-four casualties, of which nine were missing.[1] Bate reported a loss of three hundred and nineteen in his division, of which nineteen only were missing.[2] The Confederate reports as to their missing, however, cannot be reconciled with the number of prisoners turned over to our provost marshal.

The time when Bate's attack was delivered was necessarily somewhat later than the attack upon our centre and left. He was deploying and taking distance toward his left, whilst Brown was advancing, and, with the longer route necessary to reach our refused flank, he had probably a mile farther to go. This, with the halts made to correct the alignment, and to change the place of Bullock's brigade from second line to the extreme left, would account for the time which intervened between the attack of Stewart's corps east of the turnpike, and this which Bate was making on the west.

Chalmers's division of cavalry had been watching our right since noon, and being hidden from Bate's view by rolling ground, with orchards and woods, the latter thought that the cavalry had failed to accompany his movement on that flank. In fact, however, they seem to have been a little ahead of him, and some distance from his left.[3] The hollow with its brooklet, which ran diagonally across

[1] O. R., xlv. part i. p. 369. In a private letter Colonel Sherwood informs me that his missing were afterward proven to be dead. This, of course, is more or less true of all reported missing. I follow the figures of the Official Records.
[2] *Id.*, p. 743.
[3] Chalmers's Report, *Id.*, p. 764.

Ruger's front, made a turn and entered our lines in Kimball's division, between Grose's brigade (which stood next to Ruger) and Kirby's, which was Kimball's centre. Whitaker's, on the extreme right, reached to the Harpeth River.[1] The whole of Kimball's division line faced to the west, and beyond the roll of ground in that direction, in the next hollow, Chalmers prepared for his demonstration. There his horses were left with the regular details of horse holders, and his line was formed dismounted.

A road running northwest left the Carter's Creek Turnpike close to Ruger's right, and its direction was about the same as that of Ruger's breastworks where they reached the turnpike. Kimball's brigades being in echelon, his left was near a hundred yards in rear of the road, and the distance increased as one went toward the right;[2] so that opposite Kimball's centre this road was two hundred and fifty yards distant. It was there crossed by another road running out from the town through Kimball's lines, and at the intersection was the place of the reserves of the skirmish line. I have said that the whole of the 77th Pennsylvania was detailed as skirmishers for Grose's brigade, and its commandant (Colonel Rose) had his reserves near the cross-road, whilst his line of sentinels was nearly half a mile out from our main line, connecting at the Carter's Creek Turnpike with those of Ruger's division.[3] On his right he connected with the skirmishers of Kimball's other brigades.

Bate's advance pushed back our skirmish lines, and as Rose's men retreated they rallied on the

---

[1] See *ante*, p. 62.
[2] See Major Twining's Map, p. 45, *ante*.
[3] Grose's Report, O. R., xlv. part i. p. 208; Rose's, *Id.*, p. 226.

reserve farthest toward the right,[1] and took advantage of the little ravine to check the enemy from this cover. Here they held on for some time, till the advance of Chalmers's dismounted cavalry subjected them to a sharp fire on their right flank. They then retired within the lines by way of the hollow.

This movement of the skirmishers uncovered, at first, only the Pennsylvania battery and the left of Grose's brigade. These were brought into action, and the check given the enemy by the stout resistance of the picket was turned into defeat by the fire of the line. Grose says that his firing began when the enemy was about two hundred and fifty yards away, and that few of them got nearer than a hundred yards. It was a short, sharp combat, and the Confederate infantry retreated to the cover of the hill at the Bostick house.[2]

It must be remembered that only two regiments of Bate's division were deployed west of the Carter's Creek Turnpike, and they did not extend over the whole of Grose's front. This is explicitly stated in the report of the 9th Indiana (Colonel Suman), which was his right regiment, and was separated from the rest by the little ravine and watercourse. "Only a skirmish line," Colonel Suman says, "showed itself in front of my regiment, though their line of battle was seen very close in front of the 75th Illinois.[3] One volley, an oblique fire, was all the fighting my regiment did. There were no casualties." The report of the 75th confirms this, telling us that the first volley checked the enemy as soon as he was within good musket range, and a few more sent him back in confusion. This regiment also suffered no

[1] Grose's Report, O. R., xlv. part i. p. 208; Rose's, *Id.*, p. 226.
[2] O. R., xlv. part i. p. 208.   [3] Suman's Report, *Id.*, p. 221.

casualties.[1] Indeed, the casualties in the whole brigade may be said to be trifling, for out of an aggregate of thirty-seven, full half were in the 77th Pennsylvania, which was the skirmish line.[2] The regimental reports show that the firing lasted ten or fifteen minutes, when the enemy retreated and made no further serious effort on Grose's front.

As we have passed beyond the scope of Bate's attack in our examination of events as they occurred from left to right, we must conclude that whatever fighting there was on the front of Kirby and Whitaker (the centre and right brigades of Kimball's division) was with the Confederate cavalry under Chalmers. Though no fully authenticated report by this officer has been found among the Confederate archives which came into possession of the government at the close of the war, the original draught found among General Chalmers's papers and furnished by him has been accepted without question, and is printed in the Official Records.[3] That the commanding officers on the national side do not seem to have recognized them as cavalry does not militate against the facts, which prove beyond question Chalmers's advance on this part of the field, and that Bate's infantry fell far short of reaching it.

The gray uniforms of the enemy, soiled and worn with hard service, were indistinguishable in the different corps, and as to arms or equipment the case was nearly the same. Forrest relied upon the carbines or muskets of his men, and despised the sabre, so that his horsemen were more than any others a mounted infantry at a time when the usual custom

---

[1] Bennett's Report, O. R., xlv. part i. p. 214.
[2] Grose's Report, *Id.*, p. 209.
[3] *Id.*, p. 763, and *ante*, p. 83, note.

on both sides was for this arm of the service to fight in line dismounted.

The heavy firing had been heard for some time on the left by Kimball's men, when at last, about sunset, a line was seen on the ridge in front of Kirby and Whitaker.[1] Two or three volleys drove the enemy back, the skirmishers were again advanced, and though a lively skirmish fire was kept up late in the evening, they maintained their position without difficulty, and the main line in this part of the field was not again engaged.

In Whitaker's brigade the efforts of the enemy to test the strength with which the extreme flank was held were possibly more persistent than in front of Kirby, for Whitaker speaks of the fight as "short but severe. The enemy persisted in the assault about fifteen minutes, when they broke and fled, to return no more."[2] Whitaker's casualties were only one killed and eight wounded. Kirby's aggregate was fourteen, but McDanald's regiment, which was sent to reinforce Ruger's division, belonged to this brigade.

Chalmers tells us that his line advanced about half past four, driving in the skirmishers, till he was "within sixty yards of the fortifications," but his "force was too small to justify an attempt to storm them." He reports his loss at one hundred and sixteen, killed and wounded.[3] His claim to have "held his position" in Kimball's front can only mean the position behind the rolling ground, out of range, for certainly his men were seen no more from Kimball's line, nor did they manifest their presence when the troops were withdrawn at mid-

[1] Kirby's Report, O. R., xlv. part i. p. 184.
[2] *Id.*, p. 195.   [3] *Id.*, p. 764.

night. On this part of the line, therefore, we must conclude that a forced reconnoissance was all that Chalmers really attempted, and that, though boldly pushed, it was easily repulsed.

The extent of front covered by Kimball was so great that we had not attempted to make a continuous line of infantry trench, but only to hold the salient points, putting in the brigades where they would mutually support one another, and sweep with their fire the spaces between. Even this had not been fully accomplished when the fighting began, and the reports of brigade and regimental commanders show that the works were incomplete, and the troops had to drop the intrenching tools to seize their rifles.[1] This was what led both Hood himself and Bate to express their regret that a more vigorous attack was not made on our right flank. I have already given reasons for my opinion that it was not practicable, and that we were right in considering the Carter's Creek Turnpike as the limit of our proper front on that side, Kimball's division being more in the position of a support for the flank than as part of the line itself.[2]

[1] O. R., xlv. part i. pp. 195, 214. [2] *Ante*, p. 60.

CHAPTER X

THE SITUATION AT SUNSET

Determined Fighting at the Centre — Examination of Strickland's Line — Enemy holding outside of his Works — The Second Line — Relative Position of Opdycke and Strickland — Orders to the latter — Visit to Ruger — Wagner reorganizing — Visit to extreme Left — Reinforcements for the Centre.

WE have now followed the progress of the battle during the hour between four and five o'clock, and have seen that the severe repulse of the enemy on both our flanks was complete and final. In these parts of the field skirmishers had been advanced as the Confederates retreated, and were kept out till our final withdrawal at midnight. The skirmishing was active, and a lively, rattling fire continued along the front on both wings, but no serious efforts to advance again in force were made by the enemy anywhere but at and near the centre. Whenever a new assault was made near the Columbia road, brisk demonstrations would be made by Stewart's corps on our left, and by Bate's division on our right; the artillery in our line would sweep the front with canister, and the sounds would indicate a battle, but the only really determined work by the Confederates was done upon the front of the central brigades of Reilly and Strickland, between which Opdycke's sturdy regiments were now solidly placed, the lines of all three brigades being more or less thickened by disorganized but brave groups of Conrad's and Lane's men, who had rallied at the works.

My personal observation of these earlier scenes in the fight has been brought down to the time when the disorder at the centre was corrected by the splendid charges of our reserves. It may be well to take up the story there, and tell what passed under my own eye, so far as it is material to an accurate understanding of the persistent struggle.

I have said that from the turnpike we could not see what was occurring behind the Carter house and its group of farm buildings. The comfortable farmstead was surrounded by shade trees, with some fruit trees in the enclosure nearest the house. This grove was mostly on the northern slope toward the town, after the locust trees at and in front of the breastworks had been felled, and, with the house and farm buildings, quite shut off the view looking westward. As soon as the first repulse of the enemy on Reilly's line and on the turnpike left me free to turn my attention elsewhere, I went in person to examine the condition of affairs at the centre of Strickland's brigade line. I had seen Opdycke's men at the retrenchment crossing the road in line with the brick smoke-house and adjacent building in rear of the opening in the works left for the incoming troops and trains, but I could not see how they connected on the right with Strickland's men.[1] The smoke obscured the view, but enough could be seen to make me suspect that the enemy still held the outside of our main intrenchments for some distance west of the turnpike and southwest of the Carter buildings.

Passing round the grove and houses under cover of

[1] *Ante*, pp. 97, 98.

The Situation at Sunset 143

the southern slope, I saw, as soon as I got beyond the obstructions to the view, that my suspicion was correct, and that our men were holding a new line, apparently running from the brick smoke-house above mentioned, standing south of the dwelling-house.[1] The efforts of the enemy to break through at this point were so quickly renewed that the lulls were short and there was small chance for extended investigation. Not meeting Colonel Strickland in person, I sent to him, by a staff officer, an order to make the most strenuous efforts to carry his brigade forward to their original place in the front line of intrenchment.

We have seen how embrasures had been made for the four guns of the 20th Ohio Battery in the second line just west of the smoke-house and the other small building that stood between it and the road. The infantry now made a continuous line from the retrenchment in the road to and between these buildings, and among the guns. Just west of the battery was the breastwork which the 44th Missouri had built. When the break occurred at the front line, the buildings and this breastwork made a rallying place, and Opdycke's brigade on the left, with Strickland's on the right, had formed upon it as the reserves rushed forward. In the first mêlée, officers and men of both brigades were intermixed, and scattered among them here and there were some of those whose pell-mell rush from the front had swept from the intrenchment part of its defenders. The bulk of Opdycke's, however, were nearer the turnpike, and most of Strickland's gathered toward the right,[2] so that, although they lapped upon each other, Strickland occupied substantially the right

[1] See Sketch Map, *ante*, p. 43.   [2] *Ante*, p. 117.

half of the line originally allotted to him. Whenever a lull occurred in the fight, the men instinctively crowded to left and right so as to get the organizations better separated, and both brigades strengthened and extended the barricade in the second line on which they had rallied.[1]

In reply to the order to Strickland to reoccupy the front line of breastworks, he soon sent word that he had done so; but still noticing from the turnpike that, in the repeated onslaughts of the Confederates that followed each other quickly, their fire came from what I recognized as the outside of our original intrenchments, I went a second time to his position, just before sunset. I found him just west of the ell or wing of the Carter house, in rear of what was the centre of his brigade line before the battle opened, but now was near the point where his left and Opdycke's right appeared to join.

From this position I pointed out to him, as well as the smoke would permit, the difference between the present line held by his men and that which they had originally occupied, and from the other side of which the enemy were firing. Urging him to watch for an opportunity to get forward and drive his opponents away from the works, I passed on to Moore's brigade, and visited General Ruger, the division commander. The whole of Moore's front was firmly held, and Ruger's report indicated that the fighting beyond the Cartersville Turnpike had not been serious. I therefore felt free to call upon Kimball to send a regiment to Ruger to strengthen the flank of Moore's brigade and assist in Strickland's movement forward, as has been stated.[2]

In this personal visit to the different brigades in

[1] O. R., xlv. part i. p. 354. [2] *Ante*, p. 134.

line, my object was to verify the actual situation and come into touch with subordinate commanders, so that the action of all might be co-ordinated to secure victory. I now knew that Kimball and his brigades were firmly in place. So was Ruger with his, except the change in Strickland's line which has been described. Opdycke's brigade was astride the Columbia Turnpike, crowding upon Reilly on the left and upon Strickland on the right. Had there been any other division or brigade commander there, my quest would have discovered him, for my ride was an inspection for the express purpose of knowing what strength we had in hand to complete our task. I did not find Wagner or either of his brigade commanders except Opdycke. I have never doubted that the other two were where duty called them, busy in the reorganization of the two disordered brigades; but they were not in the line nor visible from it.[1]

Returning to the centre, I found that Strickland thought he needed the help of fresher well organized troops to lead forward to the breastworks in front, and I determined to continue my ride to the extreme left and judge, after conference with Reilly and his brigade commanders, whether a reinforcement for Strickland could be spared.

At my headquarters on the turnpike I was met by a message from my acting Quartermaster, Captain Hentig,[2] complaining that the passage of wagons and ambulances at the bridge was obstructed by crowds of Wagner's men. I sent my Adjutant Gen-

[1] The reason for making this statement so detailed and explicit will appear when, in chapter xix., I shall have to notice the controversies in regard to this part of the battle history.

[2] Captain Hentig was Commissary of Subsistence, but in the absence of my Quartermaster on leave, he was performing temporarily the duties of both offices.

eral to the place to ask General Wagner to correct the evil. Captain Cox found Wagner seeking to reorganize his brigades in the open space along the river, and exhorting those whom he and his officers had there halted and brought into some kind of order.[1] He promised to place a guard at the bridge to stop stragglers and keep the passage free, and this he did.

After communicating thus with Wagner, I passed on to the left and saw Henderson and Stiles in person, and, from their reports, feeling assured as to the safety of that part of the line, directed that their reserve regiment, the 112th Illinois, should be sent to the right centre to assist Strickland in fully regaining the works. My Inspector General, Major Dow, was with me, and as he was an officer of this regiment, he was, at his own request, allowed to accompany it in its new task.[2]

On my way back I consulted with General Reilly and Colonel Casement, and informed them of my action. I also directed General Reilly to have a detachment in readiness to form on the outside of his breastworks, and to sweep with an enfilading fire the enemy who held on in front of the right centre, if the new effort from Strickland's line should not succeed.[3]

[1] Captain Cox's statement, Appendix F.

[2] Henderson, the brigade commander, was colonel of the regiment, which was therefore in command of Lieut. Colonel Bond. Colonel Henderson for many years represented his district in Congress, and is one of the most honored public men of western Illinois. His recollection of my visit to the brigade for the purpose noted is included in an interesting written statement from which I have quoted, and shall have other occasions to quote.

[3] In a letter from General Casement (in 1881) replying to inquiries of mine when I was preparing the volume of history of this campaign, he mentions the last order, reminding me of an amusing personal occur-

The sun was just setting behind the hills in the west as I got back to my station on the Columbia Turnpike, close to the Carter house. Twilight and darkness soon settled down upon the field, for the true sunset was at almost exactly five o'clock.[1] Subordinates were notified where to find me, and to make their reports during the evening, so that there might be no miscarriage of communications in the darkness. The 112th Illinois had more than half a mile to march, in order to reach and reinforce Strickland, and it was completely dark when the regiment reported to him. Meanwhile the fight at the centre had been bitter and unceasing, for Hood's brave men were still possessed with the belief that they could turn their brief advantage there into a complete victory, and they fought with almost unexampled tenacity to accomplish their desire.

---

rence connected with it. In a subsequent fuller statement he confirmed it, with other matters to which I shall refer later. General Casement was colonel of the 103d Ohio, and was brevetted Brigadier General for services in this campaign. He afterward won a national reputation in connection with the track-laying of the Union Pacific Railway, and in other great works of railway construction. He was Territorial Delegate in Congress, and filled other important public positions. Few business men have been so prominent or so well known.

[1] Professor J. G. Porter of the Cincinnati Observatory has kindly computed the actual sunset at Franklin on November 30, 1864, and finds it to have been at 4 h. 59 m., local time. Colonel M. B. Carter, living on the field, has also been good enough to note the apparent sunset on one of the anniversaries of the battle, and found it to be 4 h. 51 m. The low hills westward account for the difference between the real and the apparent sinking of the sun below the horizon. He also noted that at 5 h. 2 m. it was too dark to read ordinary newspaper print, marking the twilight which is commonly called "early candle-lighting."

CHAPTER XI

FROM THE CONFEDERATE STANDPOINT

Multiplication of Lines of Attack — How caused — Walthall's Description of the Assault — In the Abattis — Repulsed in Confusion — The Ditch at the Cotton-gin — Brown's Attack on Right Centre — Bate overlaps him — Johnson's Attack after Dark — Hood's Description — S. D. Lee's — Colonel Capers's — Capture of Gordon — His Account of the Charge — Cleburne falls.

It was the impression made on those who fought near our centre, that the enemy's original formation was that of deep columns of attack, charging one after another to the number of a dozen or more, and intended to break through our lines at the Carter house by their numerous and rapidly repeated assaults. This was a natural conclusion from the manner in which the charges were renewed and persisted in, long after the flanks at right and left were relieved from serious fighting. This was, however, the result of fortuitous circumstances and not of a plan, for the official reports of all the Confederate officers prove that their first deployment was, like our own, a line of battle in which each division was arrayed with two deployed brigades in first line, supported by one in second line. The exceptions were in French's division, which (as one of the brigades was absent) had but one in front and one in support, and in Brown's division, which had four brigades, and therefore put two in each line. Reasons have already been given for concluding that Hood's real design was to turn or break our left flank, which was by far closer to the bridges and ford which were our line of retreat in case of disaster.[1]

As Hood, however, in his first advance, sent eighteen brigades to attack our main line of five between the railroad and Carter's Creek Turnpike, it necessarily turned out that the Confederate movement was a convergent one. Their brigades being the tactical units which actually controlled, these not only lapped over each other, but unforeseen circumstances made the multiplication of the brigade lines greater at the centre by the spontaneous crowding in from the flanks when the word passed among them that they had possession of our parapet near the Carter house.

As each division had its own support, there was a double assault when it struck our front, with as many successive rallyings and fresh attacks as its men could be induced to make. The divisions of Loring, Walthall, French, and Cleburne were all east of the Columbia Turnpike, and Cleburne's left guided upon the road.[2] Loring's right was necessarily guided by the line of the river and railroad, and as, by reason of the check at the centre caused by the brief resistance of Wagner's two brigades, the enemy's right got forward fastest, Walthall came in echelon to Loring, lapping over him in rear to the extent of a brigade, whilst French would be similarly in rear of Walthall.

By the same crowding process Cleburne's division would be thrown behind French, and immediately upon the road and on its east side there would result a column of five brigades, for Walthall distinctly declares that his left brigade (Shelley's) assaulted our works "just to the right of the pike," which brings him in

[1] *Ante*, p. 87.   [2] See Stewart's Map, facing p. 87.

front of Reilly's brigade on our side.[1] He further says that General Shelley was among those that got over the works; and as we certainly know that Casement's brigade was nowhere pierced, it seems clearly proved that it was between the cotton-gin and the turnpike that Shelley was in person. Reynolds's brigade had apparently been thrown into the second line by obstructions in the advance, which have been before mentioned, and it may have been in rear of Quarles, who attacked Casement.[2] But behind Shelley must have been Cockrell's and Sears's brigades of French's division, and two of Cleburne's three, making five in column. But even this does not complete the full tale; for Gordon's brigade of Brown's division got also astride of the road, and Gordon himself was captured in Reilly's brigade line, as I shall show.[3] Thus, at brief intervals, six distinct lines here made furious attacks, to say nothing of repeated rallyings and new efforts of each.

Let us summarize these attacks upon our left wing. Loring's division in two lines attacked Stiles's brigade, and made the brave but fearfully costly effort to turn our flank by way of the railway cut. Featherston's brigade of the same division seems to have come first against Casement's works, and to have been followed by Quarles's and Reynolds's brigades of Walthall's division in succession. The front of Reilly's brigade, including the Kentucky battery at the turnpike, was struck by the six brigades in succession that have been enumerated in the last paragraph.

Walthall's own report of the advance of his division is such indisputable evidence of the desperate nature of the struggle that it deserves to be quoted at some

[1] O. R., xlv. part i. p. 720.   [2] *Ante*, p. 123.
[3] *Post*, p. 159.

length. "Both officers and men," he says, "seemed fully alive to the importance of beating the enemy here at any cost, and the line moved steadily forward until it neared his outer works,[1] and then fell upon it so impetuously that the opposing force gave way without even retarding the advance, and retired in disorder to the strong intrenchments in the rear. There was an extensive open and almost unbroken plain between the outer and inner lines, across which we must pass to reach the latter. This was done under far the most deadly fire of small arms and artillery that I have ever seen troops subjected to. Terribly torn at every step by an oblique fire from a battery advantageously posted at the enemy's left, no less than by the destructive fire in front, the line moved on and did not falter till, just to the right of the pike, it reached the abattis fronting the works. Over this no organized force could go, and here the main body of my command, both front line and reserve, was repulsed in confusion; but over this obstacle, impassable for a solid line, many officers and men (among the former Brigadier General Shelley) made their way, and some crossing the ditch in its rear were captured, and others killed and wounded in the effort to mount the embankment. Numbers of every brigade gained the ditch and there continued the struggle, with but the earthwork separating them from the enemy, until late in the night."[2]

The gallant division commander does not tell of his own experience, but his corps commander, General Stewart, reports that "Major General Walthall had two horses killed, and was himself severely bruised."[3] General Quarles was severely wounded in the advance, all his staff officers with him on the

[1] The flank of Conrad's brigade at its outpost position.
[2] O. R., xlv. part i. pp. 720, 721.   [3] *Id.*, p. 708.

field were killed, and we are told that "so heavy were the losses in his command that when the battle ended its officer highest in rank was a captain."[1] It is greatly to be regretted that so few of the Confederate reports of the battle are preserved, if indeed they were made; but this of Walthall shows what must have been their character.

On the west of the turnpike Brown's division went forward in two lines. In the first line were Gordon's brigade (guiding on the Columbia Turnpike) with Gist's on Gordon's left. In the second line Strahl's was in support of Gordon's and Carter's in support of Gist's.[2] It also appears that Cockrell's brigade of French's division lapped over Gordon and Strahl. Later in the fight Bate's division lapped upon Brown's to the extent of two brigades, for he attacked with the three in single line, and his left brigade was astride the Carter's Creek Turnpike.[3] Two and a half of his brigades were therefore in front of our two (Strickland's and Moore's), and Bate's right must have reached nearly or quite to the Columbia road, making the third line which attacked our right centre. But this was not all. An hour after dark Johnson's division of Lee's corps, which was in reserve, went to the assistance of Cheatham's men, and this strong division of four brigades made a fifth and a sixth line of battle attacking near the Carter house,[4] where Opdycke was now filling half of Strickland's line and Moore's left was holding grimly to its breastwork, the regiment from Stiles on our extreme left and another from Kimball on our extreme right gallantly helping the defence at the critical point.

For the attack by Johnson's division we are depend-

[1] O. R., xlv. part i. p. 721.   [3] Bate's Report, *Id.*, p. 743.
[2] Capers's Report, *Id.*, p. 736.   [4] Lee's Report, *Id.*, p. 687.

ent upon the official report of General Stephen D. Lee, the corps commander. "This division," he says, "moved against the enemy's breastworks under a heavy fire of artillery and musketry, gallantly driving the enemy from portions of his line. The brigades of Sharp and Brantley (Mississippians) and of Deas (Alabamians) particularly distinguished themselves. Their dead were mostly in the trenches and on the works of the enemy, where they nobly fell in a desperate hand to hand conflict. Sharp captured three stand of colors. Brantley was exposed to a severe enfilading fire. These noble brigades never faltered in this terrible night struggle. Brigadier General Manigault, commanding a brigade of Alabamians and South Carolinians, was severely wounded in this engagement while gallantly leading his troops to the fight, and of his two successors in command, Colonel Shaw was killed and Colonel Davis wounded. I have never seen greater evidence of gallantry than was displayed by this division under the command of that admirable and gallant soldier, Major General Edward Johnson."[1]

Here, as on the other side of the turnpike, the fearful destruction of Confederate officers is the reason for the scarcity of official reports; but all that we have are so telling in their references to the desperation of the conflict, that we would perhaps shrink from further repetition. Hood himself says of it that "the engagement was of the fiercest possible character. Our men possessed themselves of the exterior of the works while the enemy held the interior. Many of our men were killed entirely inside the works. The brave men captured were taken inside his works in the edge of the town. The struggle lasted till near midnight, when the enemy abandoned his works and

[1] O. R., xlv. part i. pp. 687, 688.

crossed the river, leaving his dead and wounded in our possession. Never did troops fight more gallantly." [1]

There are errors in these statements, but they fairly describe the general character of the fight. Lieut. General S. D. Lee tells also how he was ordered to put Johnson's division in, because it was a stubborn fight and Cheatham had informed him about dark that assistance was needed at once. "Owing to the darkness and want of information as to the locality," he says, "his attack was not felt by the enemy till about one hour after dark." He calls it a "desperate hand to hand conflict," and a "terrible night struggle." Of our side he says, "The enemy fought gallantly and obstinately at Franklin, and the position he held was, for infantry defence, one of the best I have ever seen." [2]

Major General Clayton, of the same corps, says that his division reached Franklin late in the afternoon. "We found that bloody and disastrous engagement begun, and were put in position to attack, but night mercifully interposed to save us from the terrible scourge which our brave companions had suffered." [3]

In General Brown's division of Cheatham's corps, the report of Colonel Capers of the 24th South Carolina, from which I have already quoted,[4] is the only one preserved. He tells how his brigade commander, General Gist, had first his horse shot under him, and then, leading his brigade on foot, fell, pierced through the heart. At the locust abattis the brigade was checked, and there they captured and sent to the rear many of Wagner's men. "Fortunately for us," he says, "the fire of the enemy slackened to let their advance troops come in, and we took advantage of it to work

[1] O. R., xlv. part i. p. 653.   [3] Id., p. 697.
[2] Id., p. 689.   [4] Ante, p. 93.

our way through. Gist's and Gordon's brigades charged on and reached the ditch, mounted the work, and met the enemy in close combat. The colors of the 24th were planted and defended on the parapet, and the enemy retired in our front some distance, but soon rallied and came back in turn to charge us. He never succeeded in retaking the line we held. About dusk there was a lull in the firing west of the pike. Brown's division had established itself in the ditch of the work, and so far as Gist's brigade front, on the crest. Torn and exhausted, deprived of every general officer and nearly every field officer, the division had only strength enough left to hold its position. Strahl's and Carter's brigades came gallantly to the assistance of Gist's and Gordon's, but the enemy's fire from the houses in rear of the line, and from his reserves thrown rapidly forward, and from guns posted on the far side of the river so as to enfilade the field, tore their line to pieces before it reached the locust abattis." [1]

Colonel Capers was wounded and fell in the early part of the attack, and he does not claim to narrate the whole engagement as an eyewitness, but completed his report from statements of his subordinates.[2] It also contains errors, but it is contemporaneous and important testimony. He informs us that, at the close of the engagement, Captain Gillis of the 46th Georgia was the senior officer of the brigade, and that the men in the ditch loaded the muskets and passed them up to those who could fire over the parapet, and thus an effective fire was maintained until nine o'clock.

The meagre Confederate official reports are supplemented by recollections of Southern officers published

---

[1] O. R., xlv. part i. p. 736.
[2] Colonel Capers became a minister of the Protestant Episcopal Church after the war, and is now (1896) Bishop of South Carolina.

since the war. General George W. Gordon, who was made prisoner in the engagement, delivered in 1891 an address at the unveiling of a statue of General Cleburne, full of valuable matter drawn from his own experience in the battle. He describes the formation of Hood's army on either side of the Columbia Turnpike, Cleburne's and Brown's divisions of Cheatham's corps advancing, the one on the east and the other on the west of the highway, which was the left guide of Cleburne and the right guide of Brown. General Granbury's brigade was Cleburne's left wing, and Gordon's own brigade the right of Brown's. These two brigades moved to the attack in the front line separated only by the road. "As the array," he continues, "with a front of two miles or more in length, moved steadily down the heights and into the valley below with flying banners, beating drums, and bristling guns, it presented a scene of the most imposing grandeur and magnificence. When we had arrived within about four hundred paces of the enemy's advanced line of intrenchments,[1] our columns were halted and deployed into two lines of battle preparatory to the charge. This advanced position of the enemy was not a continuous, but a detached line, manned by two brigades, and situated about six hundred paces in front of his main line of formidable works, and was immediately in front of Cleburne's left and Cheatham's (Brown's) right. When all was ready the charge was ordered. With a wild shout, we dashed forward upon this line. The enemy delivered one volley at our rushing ranks, and precipitately fled for refuge to his main and rear line. The shout was raised, 'Go into the works with them.' This cry was taken up and vociferated from a thousand throats as we rushed on

[1] He here refers to the outpost line of Wagner's two brigades.

after the flying forces we had routed, — killing some in our running fire, and capturing others who were slow of foot, — sustaining but small losses ourselves until we arrived within about one hundred paces of their main line and stronghold, when it seemed to me that Hell itself had exploded in our faces. The enemy had thus long reserved their fire for the safety of their routed comrades who were flying to them for protection, and who were just in front of and mingled with the pursuing Confederates. When it became no longer safe for themselves to reserve their fire, they opened upon us (regardless of their own men who were mingled with us) such a hailstorm of shot and shell, musketry and canister, that the very atmosphere was hideous with the shrieks of the messengers of death. The booming of cannon, the bursting of bombs, the rattle of musketry, the shrieking of shells, the whizzing of bullets, the shouting of hosts, and the falling of men in their struggle for victory, all made a scene of surpassing terror and awful grandeur."

I have not felt at liberty to abridge this description by General Gordon, for nothing short of the whole would quite convey the impression made on his mind by that terrible charge. It is not the story of an imaginative writer of fiction, but the words of a brave soldier at the head of his brigade, giving the best and truest narrative he can of the actual situation of his command and of himself. He had at an exceptionally early age won his rank by distinguished services on many a hard fought field, and this only makes more terribly significant the pre-eminence he gives to the field of Franklin over all others he had seen. It was in the centre of this storm of war that Cleburne appears, whose memory his friend was celebrating.

"Amid this scene General Cleburne came charging down our line to the left, and diagonally toward the enemy's works, his horse running at full speed, and if I had not personally checked my pace as I ran on foot, he would have plunged over and trampled me to the earth. On he dashed, but for an instant longer, when rider and horse both fell, pierced with many bullets, within a few paces of the enemy's works. On we rushed, his men of Granbury's brigade and mine having mingled as we closed on the line, until we reached the enemy's works; but being now so exhausted and so few in numbers, we halted in the ditch on the outside of the breastworks among dead and dying men, both Federals and Confederates. A few charged over, but were clubbed down with muskets or pierced with bayonets. For some time we fought them across the breastworks, both sides lying low and putting their guns under the head-logs upon the works, firing rapidly and at random and not exposing any part of the body except the hand that fired the gun." Suffering from fire in every direction, — front, flanks, and rear, — they finally shouted to our men within the works that they would surrender. "At length," says General Gordon, "they heard us and understood us, ceased their fire, and we crossed their works and surrendered. It was fatal to leave the ditch and endeavor to escape to the rear. Every man who attempted it (and a number did) was at once exposed and was shot down without exception." He concludes his description with an important fact which fixes his own location as well as that of the rest of his brigade.

"The left of my brigade," he says, "under command of Colonel Horace Rice (I was on the right) successfully broke the line, and some of my brave and noble men were killed fifty paces or more within the works. But just at this critical juncture a reinforcement of a Federal brigade confronted them with a heavy fire, and being few in numbers they were driven back to the opposite side of the works, behind which they took position and bravely held the line they had previously taken."[1]

The particulars so vividly told by General Gordon leave no doubt as to his position on the field. The left of his brigade is placed at the break in Strickland's line on the west of the turnpike, whilst his right (with which he was) extended across the road to the east, occupying part of Cleburne's ground, and mingling with Granbury's brigade. This explains also Cleburne's hasty ride towards them to see the cause of the interference in the movement, and corroborates the general opinion that that distinguished officer fell on the east of the turnpike, nearly in front of the Carter cotton-gin and a very few rods from our breastworks.[2] It shows also that Gordon and his surviving companions came over the works as prisoners in Reilly's brigade line not far from the cotton-gin, the fire from that salient being mentioned as one of the causes of his surrender, and the head-logs spoken of being east of the Kentucky battery front, which was constructed with embrasures for the cannon, and not with the head-log for infantry protection.

[1] The address from which these extracts were taken was delivered at Helena, Arkansas, and published in the Memphis Appeal of May 11, 1891.

[2] See also a statement by Mr. John McQuaide of Vicksburg, one of the party which found the body. Century War Book, iv. 439. It substantially agrees also with the position shown by the Carter family and other residents of the neighborhood.

CHAPTER XII

THE BATTLE AFTER DARK

In the Locust Grove — The Two Lines — The 112th Illinois — Reilly's Detachment — Sweeping the Ditches — Captain Cunningham's Story — The Sergeant Major's — General Strahl's Death — Hood's Reserves — Rallying on them — Later Alarms — Orders to Wood's Division — Preparations for Withdrawal.

HAVING seen from the Confederate standpoint the repeated heroic assaults made upon our lines, we must again turn our attention to the progress of the equally determined defence, especially at the critical point at the locust grove, where our men had given back from the main line and were holding their second line of hastily constructed barricade. It will be remembered that, after the main line of works crossed the Columbia Turnpike and ran west some fifty yards, it turned toward the northwest, following the curve of the hill, and descending a little upon its slope.[1] The body of the enemy that held on to the outside of the parapet and filled the ditch was just west of the angle in the line where the men were sheltered from the fire of the salient near the cotton-gin by both the angle itself and the depression of the ground below the level of the turnpike. The Confederates there were hardly an organized body, but were a mingling of men from all the different commands that had followed Gordon's and

---

[1] See *ante*, p. 43, and the sketch there given.

The Battle after Dark 161

Gist's brigades in the first charge. In this respect the confusion of commands on both sides was not dissimilar. Those of the enemy who were within the range of our oblique fire from left or right were swept away, and as to the crowd at the ditch at the place indicated, it may fairly be said that it was less perilous for them to remain than to retreat across the storm-swept field behind them. There they stood in desperation, the men in the ditch passing up loaded guns to those who stood on the berme, who were thus able to keep up a continuous fire over the parapet.

It was desirable that the fire from the salient at the cotton-gin should have the most effective enfilading sweep along the front, and it was for this reason that the 112th Illinois, when brought from our left, was ordered to report to Strickland for the purpose of bringing our line forward to the main works by advancing from his right. Lieut. Colonel Bond, commanding the regiment, accompanied by Major Dow of my staff, led it to the right of Strickland's brigade,[1] the position of the 72d Illinois, which was, by the casualties of the battle, left in command of Captain Sexton. This intrepid officer was consulted as to the extent of the enemy's front in contact with the earthworks, and it was decided to pass Colonel Bond's regiment out near the small log building or corn-crib which was standing in rear of Sexton's line,[2] and try to reach the main works close to the flank of Moore's brigade, supporting the movement by the troops in second line advancing on the left of Bond as he reached the parapet. Captain Carter of the 72d Illinois acted as guide for Colonel Bond in the advance.

[1] *Ante*, p. 146.   [2] See sketch, *ante*, p. 43.

As it had already become dark, in the double gloom of night and the pall of smoke the position and the extent of the enemy could only be known by the flashing of the musketry. The 112th Illinois got over the barricade in front of Sexton's line and crawled forward, led by Bond and Dow, and keeping low so as to be beneath the line of fire. They had no great difficulty in getting close to the parapet in front, but when they rose to occupy the work the fire of the enemy was so close that, as Colonel Bond says, his face was burned by the powder.[1] But the fire in front was not the only peril in the darkness. Although word had been sent to the troops on right and left to cease firing, the din of battle made it hard to get the orders understood by the men in line, and they could not be restrained from firing obliquely at the flash of the enemy's guns. Sexton's regiment had gone forward on the left of Bond's, and both dropped to the ground to wait for a cessation of the fire. Colonel Bond was twice wounded slightly, one ball cutting the tendon of his heel. "As I was very close to the first line," he says, "the Confederates could not have fired so low."[2]

Finding that the front line was now made untenable by this fire from their friends, the regiments returned to the second line, the 112th Illinois passing through the locust grove and around the west end of the barricade, which, in Colonel Bond's judgment, extended some fifty feet west of the log building, and there ended without connecting with Moore's brigade, whose left rested on the northwest slope of

[1] Letter of Colonel Emery S. Bond, September 29, 1885. Colonel Bond was then a commission merchant in Chicago. See also statement of Major Dow, in Appendix E.
[2] Letter of October 22, 1885.

the Carter Hill, as Moore's report tells us.[1] The regiment remained during the evening in support of Strickland's brigade, awaiting a more favorable opportunity for the renewal of the effort to reoccupy the main works. The two lines were here, as the officers stated, about three or four rods apart.

To keep up the active efforts to clear the front of Strickland's brigade, I had directed General Reilly to put a detachment from his own brigade over the works near the cotton-gin, giving orders to sweep down along the front to the turnpike upon the flank of any of the enemy that might be hanging on in front of Opdycke, thus assisting in dislodging those also who were farther down behind the angle already described.[2] This he did, the result being considerable additions to the number of prisoners and flags in our hands. It was now near nine o'clock in the evening. The attack of General Edward Johnson's division from Hood's reserve had been made and repulsed, and as it fell back from the hollow toward the higher ground, where the barricades held by Wagner's men at the beginning of the fight still were, the remnant of the Confederates in the ditch who were not too much disabled made their way also to the rear. Heavy volleying and continuous fire was still for some time kept up, but it was now across a considerable interval, and not from the outside of our own parapet at any point.

[1] Captain Sexton fully corroborates Colonel Bond's account of these details. In February, 1894, without any knowledge of Colonel Bond's letters referred to above, he wrote me an interesting statement accompanied by a sketch map from memory. At the time of writing, Captain Sexton was just closing his term as postmaster of the city of Chicago. It would be hard to find a stronger concurrence of independent testimony than that which establishes the situation on this part of the field. For Moore's Report, see O. R., xlv. part i. p. 379.

[2] *Ante*, p. 146.

Captain W. E. Cunningham, of the 41st Tennessee, in Strahl's brigade, who was himself among the severely wounded, and was among the last to crawl back from the fatal ditch in the locust grove, has described the situation clearly and vividly.[1] "The remnants of Strahl and Gordon held the works in pure desperation. It was certain death to retreat across that plain, and equally as bad to remain. The men fought doggedly across the works without officers, and with no light save the lurid glare of the enemy's artillery, which seemed to sear the eyeballs. This portion of the works was held against every attack of the enemy to regain them. The thicket had been cut down as if by a mowing machine, and the ground was all in deep furrows. About nine o'clock the firing gradually dwindled into a slight skirmish. Those who were able walked or crawled back from under the works."

Sergeant Major Cunningham, of the same regiment (a near kinsman of the captain), has also left a truthful picture of this doomed and dwindling remnant, when, as he says, there was not an efficient man left between this group and the turnpike, and among themselves hardly enough to hand up guns to the short and thin line firing from the outside of the parapet.[2] "It was evident that we could not hold out much longer, and it was thought that none of us would be left alive. It seemed expedient that we should either surrender or try and get away, when the General (Strahl) was asked, and he responded, ' Keep firing,' and just as the man to my right was

[1] Communication to the Philadelphia Times, May 27, 1882.

[2] I have referred to the reliability and value of Sergeant Major Cunningham's pamphlet in my volume on the campaign, "Franklin and Nashville," etc., p. 93.

shot and fell against me with terrible and loud groans, General Strahl was shot. He threw up his hands, falling on his face, and we thought him dead; but in asking the dying man, who still rested against my shoulder, how he was wounded, our General, not dead, and thinking my question was to him, raised up, saying he was shot in the head, and called for Colonel Stafford to turn over his command. He crawled over the dead, the ditch being three deep, about twenty feet to where Colonel Stafford was. His staff officers started to carry him to the rear, but he received another shot, and directly a third, which killed him instantly. . . . Captain W. E. Cunningham had lost an eye, and as he sat in the ditch waiting for the terrible night to end, called to his devoted comrade and told him of his misfortune. An almost helpless handful of us were left, and the writer was satisfied that our condition was not known, so he ran to the rear to report to General Brown commanding the division. He met Major Hampton of the division staff, who told him that General Brown was wounded, and that General Strahl was in command of the division. This confirmed his prediction, so he went on the hunt of General Cheatham, and after having failed to find him for some time, and seeing that relief was being sent, he lay down to rest and sleep. His shoulder was black with bruises from firing, and it seemed that no moisture was left in his system. These personal mentions are all he can give, for it was night, and the writer never knew other than what he saw. It was not long after the recruits [reserves?] were sent forward until the last gun was silenced."

Hood's reserves consisted of Clayton's and Stevenson's divisions of Lee's corps after the repulse of

Johnson's division of the same corps. Reference has already been made to the fact that they were put in position to attack.[1] General Stovall, who commanded the advanced brigade of Clayton's division, states that he occupied "the enemy's first line of works," as the Confederates designated the outpost line of Wagner's two brigades as distinguished from our "main line."[2] Just in rear of this position, sheltered by the roll of the ground, the rest of the reserve was placed, "preparatory," as General Stevenson reports, "to an assault which it was announced was to be made by the entire army at daybreak."[3] Upon this centre the remnants of the corps of Cheatham and Stewart were rallied after the repulse of Johnson's division and the return of the handful from Brown's division who escaped destruction, as has been so graphically told by the two Cunninghams.

The front of both armies was now covered by skirmishers, and, though the Confederates did not venture any further assaults, "alarms occurred frequently until eleven o'clock, and frequently caused a general musketry fire on both sides from our centre toward the right, but I found no evidence that any real attack was made at so late an hour, the demonstrations being manifestly made by the rebels to discover whether our lines were being abandoned during the evening."[4]

I have followed the successive stages of the fighting upon the line down to the time in the evening when the enemy definitely accepted defeat, and sought only to reform his lines and collect the remnants of his broken divisions under the shelter of the

[1] *Ante*, p. 154. [2] O. R., xlv. part i. p. 701. [3] *Id.*, p. 694.
[4] My own report, O. R., xlv. part i. pp. 354, 355, and Appendix B.

two divisions of Lee's corps, which had not been engaged. We could not tell, in the darkness, much beyond what was revealed by the partial contact of the skirmish lines. From time to time a general fire would roll along the front of Hood's reserve, our skirmishers would run in, and the fire would be returned from our line. After a little it would quiet down, skirmish lines would be sent forward again, and the rattle of single shots here and there would take the place of the continuous roar. When this had been a few times repeated, we became satisfied that no further attack would be made, and the enemy concluded in like manner that we were determined to hold our position through the night.

It will be remembered that General Schofield's orders in the morning had directed a withdrawal to the north side of the Harpeth at dark, if Hood had not before that time engaged us on our lines. The orders sent to Wagner on Winstead Hill prolonged his duty as rear guard in that contingency.[1] The south bank of the river would have been cleared of everything by six o'clock, and Wagner's division would have been the last to cross the Harpeth. Hood's attack had changed all that. From his position at the fort on Figuer's Hill, General Schofield could himself see the progress of the battle better than we could who were in its din and smoke, and for the details of what was occurring upon the line he had the help of his staff, who were kept constantly going and coming between his headquarters and my own at the Carter Hill.

During the desperate mêlée he had ordered General Wood's division of the Fourth Corps to be held in readiness to cover the crossing of the river[2] in case

[1] *Ante*, p. 66.   [2] O. R., xlv. part i. p. 1174.

the enemy should succeed in breaking our lines on the south side, and dispositions were carefully made by General Wood for that purpose.

A little earlier, on General Wilson's report that the enemy's infantry was threatening to cross at Hughes's Ford, three miles above Franklin, Schofield had directed a brigade from Wood's division to be sent to support the cavalry there,[1] and to delay the flank movement of Hood, which was his most probable, as it was certainly his wisest movement. General Wood ordered Beatty's brigade to that duty,[2] but it was delayed a little for the issue of rations, and before it actually marched the advance of the Confederates against us in force solved the problem of Hood's intentions, and the order was countermanded. General Wood, in accordance with the later orders, now made the disposition of his division to cover the withdrawal of our troops from the intrenched line if it should become necessary.[3] Beatty's brigade was deployed on the north bank of the river above the town; Streight's brigade along the bank of the river immediately opposite the town; and Post's brigade on the bank of the river below the town.

The telegraphic correspondence between General Schofield and General Thomas was going on at this time, and on the announcement of Hood's preparing to cross the Harpeth on our left flank above the town, Thomas ordered the army trains sent back to Brentwood and Nashville, covered and followed by the troops.[4] The stirring scenes of the battle quickly followed, necessarily suspending the orders for retiring at dark. When night shut down on the field,

[1] O. R., xlv. part i. p. 1178.   [2] *Id.*, p. 1174.
[3] Wood's Report, O. R., xlv. part i. pp. 125, 126, and Appendix C.
[4] *Ante*, p. 43, and O. R., xlv. part i. p 1171.

The Battle after Dark 169

although the conflict seemed as fierce as ever, we who were upon the line knew that the impetus of Hood's assault was broken, and that we could hold our position. Colonel Wherry, the chief of staff, had been able to report this confidence to General Schofield, who now issued new orders for a midnight withdrawal, anticipating the total cessation of the battle by that time. These directed that the lines should be drawn back from the right and left of the centre at the Carter Hill simultaneously, troops from the left passing by the railroad bridge, and those from the right by the foot-bridge, all then marching to Brentwood by the Nashville Turnpike. The skirmishers of the whole line were to be kept out until the rear of the columns had crossed the river, and were then to be withdrawn together. General Wood was to cover the crossing, as he had earlier prepared to do, and then to act as rear guard.[1]

This order was issued between six and seven o'clock, and received by me about the last named hour.[2] Fearing that General Schofield did not know the full confidence I had in our ability to hold our lines, I sent at once my Adjutant General, Captain Theodore Cox, to his headquarters to express strongly my willingness to be personally answerable for holding the position, and my opinion that there was now no need to retreat. The trains had, however, been some hours on the road toward Nashville under General Thomas's order, and with a warmly congratulatory message, General Schofield

[1] O. R., xlv. part i p. 1172.
[2] The order from Fourth Corps headquarters in obedience to General Schofield's was issued at 7.15 P. M. Allowing for time to make copies of the original, this supports my statement that General Schofield issued his between six and seven. No hour is marked on the original. *Ibid.*, and p. 1173.

sent word that it would be necessary to carry out the orders already issued.[1]

At General Schofield's headquarters my staff officer found General Stanley, who had gone there after his wound had been dressed and the necessary changes of clothing had been made. Somewhat later, my Inspector General, Major Dow, went also to the army headquarters in obedience to a call to meet there Colonel Hartsuff, Schofield's Inspector General, and to arrange the details of the order for covering our withdrawal by the skirmish lines and then bringing these off together at a later hour. Major Dow also met General Stanley at the headquarters.[2]

Another incident of the early evening was the reestablishment of the second line at the centre as it had been organized before the battle. Colonel Rousseau of the 12th Kentucky describes this in his report.[3] The repulse of the enemy at the first series of fierce assaults made it, as he says, comparatively easy to hold the works afterward, and, a skirmish line in connection with that of Casement's brigade having been advanced from the left front of Reilly's brigade, Rousseau's regiment was ordered back into the reserve position he occupied at the beginning. The right of Reilly's brigade was longer involved in the struggles which continued on the west of the turnpike.

But in Opdycke's brigade also the dictate of military judgment was obeyed, though necessarily later,

---

[1] For the detailed statement of Captain Cox, see Appendix F. This was corroborated by a letter from General Schofield to Captain L. T. Scofield, February 2, 1887, the latter having published the incident in his paper on the campaign quoted *ante*, p. 103.

[2] See also chap. xxi., *post*.

[3] O. R., xlv. part i. p. 416.

and a brigade reserve was also drawn out of the overcrowded front line, and placed in support before the movement in evacuation of the lines was begun.[1] The men of Lane's and Conrad's brigades who had remained scattered among the organized forces in the line went back to join their standards near the river, where General Wagner had been all the evening busy in the reorganization of his command.[2]

Before describing the midnight withdrawal of the troops, however, it will be proper to narrate the part in the day's work which was performed by the cavalry under General Wilson. For the sake of clearness, I have, as far as possible, omitted reference to their movements, and must go back and give their positions in the early morning, as well as their spirited engagement in the afternoon.

[1] O. R., xlv. part i. pp. 251, 253.
[2] For fuller discussion of this point, see chap. xix., *post*.

CHAPTER XIII

WILSON'S CAVALRY ENGAGEMENT

Morning Positions — Covering both Flanks of the Army — Forrest's Advance — The Fords of the Harpeth — Confederates cross at Hughes's Ford — Wilson attacks — Sharp Combat — Enemy retreat across the River — Covering the March to Nashville.

DURING the night of the 29th, the principal body of General Wilson's cavalry force was concentrated at the cross-roads two miles and a half east of Franklin, on the Triune road.[1] His headquarters were at the Matthews house, and his outposts observed the fords of the Harpeth River for several miles above the town.[2] The only important detachment was Hammond's brigade, which had been sent to Triune to learn whether Forrest was making a turning movement in force, as Wilson apprehended.[3] Hammond's brigade reported directly to the commander of the cavalry corps, and the rest of the force was organized in two divisions, Johnson's and Hatch's. The former consisted of two brigades, Croxton's and Capron's, the latter commanded by Colonel Thomas J. Harrison.[4] Under Hatch were Coon's brigade and Stewart's.

Soon after daylight of the 30th, in obedience to orders from General Schofield, Wilson sent Croxton's

---

[1] O. R., xlv. part i. p. 550.
[2] Id., p. 598.
[3] Id., pp. 1145, 1146.
[4] Id., pp. 576, 598.

brigade across the Harpeth to the south side, and thence out upon the Lewisburg Turnpike to the Douglass church.[1] Here Croxton covered the flank of our infantry column marching upon the Columbia Turnpike, and when the rear guard of the Fourth Corps passed this point Croxton also retired and formed upon General Wagner's left (facing south), Wagner having halted upon the Winstead Hill and checked the advance of Hood's army.

In pursuance of similar directions from headquarters of the army, Wilson directed Hammond's brigade not to remain at Triune if no considerable force of the enemy were found there, but to return by way of Petersburg to Wilson's mill, a position on the Brentwood Turnpike, northeast of the general position of our cavalry at the Matthews corners, and covering our communications with Nashville on the east and by the rear.[2] Hammond had found no signs of the enemy near Triune, and in the course of the forenoon took up the position indicated.[3]

To cover the right flank of the army, well out toward the west, and to try to open communication with General Cooper, who, with parts of two brigades was supposed to be marching from Centreville on the Duck River to join us at Franklin, the Fifth Iowa cavalry was detached from Harrison's brigade and sent down the river to the turnpike leading from Hillsboro to Nashville.[4]

The array of General Schofield's army was thus completed and made compact as possible. Apart from the fault inherent in a position astride of a river which circumstances had forced upon us, the preparation to meet the enemy was a satisfactory

---

[1] O. R., xlv. part i. pp. 559, 572.
[2] Id., p. 559.
[3] Id., pp. 1182, 1183.
[4] Id., p. 598.

one, especially in the relations of the cavalry to the rest of the command.

On the part of General Hood, the knowledge that Cooper was somewhere to the westward of our line of march and might at any time appear on his left flank, had made it necessary to divide Forrest's cavalry corps more nearly in equal parts, sending Chalmers with a division and a fraction of another to the Carter's Creek Turnpike, whilst Forrest himself led Buford's and Jackson's divisions in advance of the army, and endeavored by flank movements to cut in upon and destroy our trains.

When Wagner had halted at the Winstead Hill and checked there the Confederate advance, Croxton's cavalry brigade was, as we have seen, upon his left.[1] Forrest awaited the arrival of Stewart's corps, and, leaving to that officer the direct movement by the turnpike, took both his mounted divisions across to and beyond the Lewisburg road, turning the position of Opdycke's brigade on the hill, and gradually forcing Croxton back toward Franklin. Then followed the retreat of Wagner to his unfortunate position between the armies; but Forrest did not hasten the advance of his overwhelming force against Croxton's brigade. He awaited quietly the complete deployment of Hood's army, timing his advance by it. He ordered the whole of Buford's division, dismounted, to deploy on the right of Loring's division of infantry, and to go forward with it when Stewart should give the word to attack.[2] The horses were left in the wooded hollows where the deployment had been made, under cover from the fire of our artillery in the line and in Fort Granger on the hill north of the river.

[1] *Ante*, p. 173.   [2] O. R., xlv. part i. p. 754.

Jackson's division was not dismounted at once like Buford's, but was ordered to reconnoitre toward Hughes's Ford, which was nearly east from the extreme right flank of the Confederates when they passed the Winstead Hill. The road from the ford northward led directly to General Wilson's position at the Matthews place. When the general advance should be made, Jackson's directions were to cross the river and dislodge our forces from the hill at Fort Granger, from which Cockerill's rifled guns had opened fire as soon as Forrest's men had come within range.[1]

Croxton reported that about two o'clock the enemy's cavalry made a dash at him, but, being repulsed, moved off to Hughes's Ford, whilst infantry took their place.[2] The Confederate reports are so explicit in stating that Buford's dismounted men occupied the space between the Lewisburg Turnpike and the river whilst Jackson's division made the movement toward the ford, that there is little room for doubt that Croxton mistook the dismounted men for infantry, as was so frequently done when the horses and their holders were not in sight.[3] We have seen the mistake appear in the reports of Kimball's division on the extreme right, and shall

---

[1] O. R., xlv. part i. p. 754. It was the general belief among the Confederates that we had a battery upon a hill farther up the north bank of the river than the fort. This is indicated in the map accompanying Stewart's report (Official Atlas, plate lxiii. fig. 3), and is mentioned in Major Sanders's valuable paper (Southern Bivouac, July, 1885, p. 9). The belief is nevertheless a mistaken one. No artillery was with Wilson's cavalry, all the batteries of the Fourth Corps were in our main lines, and the Twenty-third Corps batteries, except that in the fort, were parked with Wood's division, and so remained during the day. O. R., xlv. part i. p. 432. (Report of Chief of Artillery, Twenty-third Corps.) For Stewart's Map, see p. 83, *ante*.

[2] *Id.*, p. 573.   [3] *Ante*, p. 138.

find it again in Forrest's report of his advance against Wilson's position on our extreme left, later in the day.

Within the limits of the field of battle there were two fords across the Harpeth. Hughes's, which has been already mentioned, was some three miles above Franklin, and about as far from the Columbia Turnpike where that road passes the Winstead Hill. McGavock's Ford was a mile and a half from the centre of the town, and not far from the Lewisburg Turnpike, which for some distance follows the general direction of the river.

Seeing the array of Hood's army in line of battle, Croxton withdrew his brigade,[1] and crossed the river to the right bank by McGavock's Ford, having one regiment (the 2d Michigan) on the hither side, which for some time skirmished with Buford's advancing men. About three o'clock, and nearly simultaneously with the advance of Hood's infantry against our lines, Wilson was definitely informed of Forrest's crossing at Hughes's Ford, and prepared to meet the enemy's movement. The first report from the outpost up the river, when Jackson's division first approached the ford, had brought news that both infantry and cavalry were threatening to cross, and, this being sent to General Schofield, he had ordered General Wood to send a brigade of infantry to support our cavalry, as we have seen.[2] But the desperate assault of Stewart's and Cheatham's corps coming at that time had made Schofield revoke his order till it should be seen whether Wood's reserve might not be more needed at the centre. It turned out that Jackson was not supported by infantry; the one division of cavalry was all that crossed to the right

[1] O. R., xlv. part i. p. 573. [2] *Ante*, p. 168.

bank of the Harpeth, and this Wilson was able to take care of. He ordered Hatch's division to advance from the Matthews place to meet the enemy in front, and Croxton to march from McGavock's Ford and attack in flank.[1] Harrison's brigade was held in reserve on the left and rear.

The Confederate cavalry advancing from Hughes's Ford had driven back the picket with its supports, Ross's brigade having the advance.[2] A series of lively charges and countercharges between single regiments occurred before Hatch could move the body of his command forward. The Confederates had gained the summits of several steep hills looking down abruptly into a little valley between them and our advancing cavalry. Their line extended beyond Hatch's left. The latter dismounted his men and moved them forward in line to the foot of the hills, when the order was given to charge. The men now sprang forward with great spirit, carrying the crests. Giving his men a moment to breathe, Hatch again ordered them forward, and the enemy broke.[3]

Jackson's men now retired across the river, the pickets were established at the fords as before, and there was no further collision between the cavalry forces on this wing of the two armies. At the time

[1] O. R., xlv. part i. pp. 560, 573.

[2] General Ross is the only subordinate of Forrest whose report of this affair is found in the Official Records. It seems to claim that his brigade was the only force which fought with Hatch and Croxton. *Id.*, p. 770. Forrest's report, however, states that Jackson's division crossed and engaged both infantry and cavalry. *Id.*, p. 754. The infantry was imaginary, and some other parts of the very general statement are inaccurate. Ross's report is the safer one to follow in matters of detail, as he was in personal contact with events on the field.

[3] O. R., xlv. part i. p. 576.

Croxton received his orders to co-operate with Hatch in the attack upon Jackson's division, he withdrew his regiment from the left bank at McGavock's Ford, but was delayed in his advance by a report that the Confederates were crossing still further down, near the town. He left two of his regiments to look after this reported movement, and marched up the stream with the remaining two. The report proved unfounded, and he recalled the regiments. Marching now with his whole brigade, he moved forward in support of Hatch's left, and took part in the pursuit of Jackson.[1] At five o'clock General Wilson was able to report the defeat of the enemy in his front, and his messenger reached General Schofield a half hour later, when our repulse of the first assaults of the infantry along the line gave him assured confidence that we should be equally successful against the reiterated efforts of the Confederates, however persistent.[2] Wilson soon followed his messenger in response to an invitation sent him just before the battle opened,[3] and early in the evening had a personal conference with General Schofield, in which he received full instructions as to his part in the general withdrawal toward Nashville. The cavalry was ordered to maintain its position till daylight, keeping its pickets at the upper fords, and then to withdraw by the Brentwood or Wilson Turnpike, being joined by Hammond's brigade as it went, and seeking to keep abreast of General Wood, whose division of infantry was to leave the river by the Nasvhille Turnpike at the same time.[4] Brentwood was indicated as the next place of concentration for the army.

[1] O. R., xlv. part i. p. 573.
[2] *Id.* p. 1179.
[3] *Id.*, p. 1178.
[4] *Id.*, pp. 1179, 1185.

On the Confederate side, Forrest withdrew Buford's division to the place where their horses had been left, and kept Jackson's division near Hughes's Ford in observation and covering the right of the army.[1] A few hours had wrought a great change in the outlook, as seen by the leader of the Confederate cavalry. When the line of battle was forming for the attack, his boiling impatience led him to ride along the infantry lines of Stewart's corps, urging haste in formation, so that the Yankees could not retreat before they could be brought to battle.[2] At midnight there was need of all his indomitable spirit to resist the conviction rife among the remnants of the shattered Confederate army, that, as an army, it was nearly destroyed, and its career practically ended.

[1] O. R., xlv. part i. p. 754.
[2] This incident was told me by a staff officer in French's division of Stewart's corps.

CHAPTER XIV

OUR WITHDRAWAL

The Medical Department — Field Hospitals — Ambulance Train — Sick and Wounded sent to Nashville — Work of Surgical Corps during the Battle — Artillery gradually withdrawn — Arrangement of Skirmish Lines — Movement of the Infantry — Orders as to Kimball's and Wagner's March — Misunderstanding — March of Ruger, Opdycke, and Reilly — A burning Building — The Field in Front — The March to Nashville.

In the preparation for evacuating our lines at midnight, the removal of the wounded from the field and the disposal of the field hospitals became an important part, and an outline of the practical organization of the medical department in a campaign should be understood.

Our surgeons did not enter upon their battle duties at Franklin without some embarrassment from the accumulation of patients during the preceding week. The ambulance train as it came in was pretty well filled with the men wounded in the affairs about Columbia and Spring Hill, and with the sick. The first wintry weather had naturally been followed by an increase of hospital cases among the troops. All these must be disposed of so that the ambulances could be ready for new work.

As soon as I assumed the temporary command of the corps, Surgeon C. S. Frink, the Surgeon in Chief of my division, took in hand the task of preparation for active work. The organization of the medical department in the field was an excellent one, the outgrowth of the experience of severe campaigns. The army and corps Medical Directors had, of course, a general supervision, but, under them, the division constituted the principal unit of administration in the field. Its medical supply train was independent, and contained the full equipment for a field hospital in tents, stores, medicines, and surgical instruments and appliances. A surgeon was selected as the head of the hospital by the Surgeon in Chief, and was detailed by proper order from headquarters. The medical officer thus in charge of the field hospital was responsible for the selection of its site, for the erection of the tents, the arrangements for operating, and the general administration which would make the hospital work run smoothly and rapidly. The hospital and ambulance train was under the charge of a lieutenant of the line, selected for his fitness for the work, and the drivers and stretcher bearers were under his military command, as he was under that of the Surgeon in Chief and the surgeon in charge of the hospital. The stretcher bearers were permanently detailed men, who were drilled in their duties and thoroughly efficient in them. The ·division medical train consisted of thirty-six ambulances and six wagons.

The Surgeon in Chief was assisted by an Operating Board of three surgeons, and as many assistant surgeons as were required, all these being detailed from the regiments on the selection of the Surgeon in Chief. Each operation of the graver class was performed under the immediate direction of one of the Operating Board. Surgeon Frink was ably seconded in the Third Division by Surgeons G. A. Collamore,

J. H. Rodgers, and C. W. McMillin of the Board, and by Surgeon Joseph S. Sparks as Chief of the Hospital. Lieutenant Alexander W. Beighle was in charge of the ambulance train. These officers worked together during the campaigns of the last two years of the war, and their harmony, discipline, system, and professional ability was of wonderful benefit to the command.

To complete the sketch of the practical organization of the surgeon's department, it is only necessary to add that in battle brigade depots for the wounded were placed as near the fighting line as practicable, in any spot sheltered from direct fire. Here the first temporary surgical assistance was given, and from these the ambulances carried the disabled to the field hospital in the rear. These depots were conducted by the surgeons and assistant surgeons of the regiments who were not detailed for the hospital work, as the distribution of labor would be made by the Surgeon in Chief in the exigencies of varying circumstances. The whole medical corps worked together with a zeal and self-devotion as worthy of remembrance as that of their comrades of the line.

When the ambulance train came in from Columbia in the forenoon, Surgeon Frink's first duty was to arrange for the proper disposal of the sick and wounded who filled it, so that it could be ready for further use. Fortunately he found a railway train of box cars on the north side of the river, loaded with forage, and just ready to start for Nashville. The surgeon persuaded the conductor of the train to wait a few minutes, while he galloped to General Schofield's headquarters, and procured from the general in person an order to unload cars enough to accommodate the disabled men of both army corps, and to take these to Nashville. The ambulances with their loads could not safely cross the river at the rough and deep ford, and waited for the repairs on the county bridge to open this way of passing to the north bank. It was thus made late in the day before the transfers could be made, and the railway train started toward its destination. Surgeon Sparks, meanwhile (acting for the Twenty-third Corps), had selected a place for the general field hospital on the Nashville Turnpike some distance north of the bridges, the tents were pitched, and the hospital organized for work. The time of Hood's delay in his attack had barely sufficed for these arrangements.[1]

In the Fourth Corps the medical organization was similar to that of the Twenty-third which I have described. The field hospital was placed on the west of the Nashville Turnpike, not far from that of the Army of the Ohio. It was in charge of Surgeon R. J. Hill of Kimball's division, under the direction of Surgeon J. T. Heard, Medical Director. The medical staff of both corps co-operated with complete harmony, and the description of the methods of one will apply to both.

The failure of the Confederate attacks upon the flanks of the position, and the consequent early cessation of severe fighting there, had made it easy to carry out the regular system at these parts of the

---

[1] I have been assisted in these details by an interesting letter from Surgeon Frink, dated June 26, 1881. Several years later he wrote for me a still more full account of the work and organization of the medical corps from which I have also borrowed. After the war he returned to his home at Elkhart, Indiana, and until his death was known as one of the eminent men of his profession and a devotee of scientific investigation.

line. The ambulances could reach the brigade depots without difficulty after the first hour's fighting,[1] and the disabled were promptly carried to the field hospitals on the north side of the river.

At the centre the situation was a much more troublesome one. The turnpike there came from the village by so gentle a slope that the road was storm-swept with rifle balls till late in the evening, and the ambulances could not approach the lines by that route. They came up the hollows on right and left as near as they could, and the wounded were carried by the stretcher bearers to meet them. Those who had fallen in the retreat of Wagner's two brigades were out in front, mingled with the Confederate dead and wounded, and could not be reached. Others lay between the two lines in Opdycke's and Strickland's brigades, and as the battle waned in the evening, were brought in and sent to the rear. But in a night engagement the darkness prevents the full scrutiny of the field by the officers, and men crept into buildings that were near, thinking only of present shelter, and forgetting that they were liable to be overlooked and left to fall into the hands of the enemy.

Some that were not wounded were so overcome by the fatigue of several days' marching and fighting that they too fell asleep wherever they might be, and did not wake till their companions had marched away in silence, the orders to "Fall in" being given with bated breath. Stragglers had taken refuge in various places in the village, secure in the darkness which prevented the provost guards from finding them. It thus happened that there were some who fell into the enemy's hands who would not have done

[1] See *ante*, p. 126.

so had the battle and the withdrawal occurred in daylight.

The same lateness of the hour prevented the surgeons at the division hospitals on the north side of the Harpeth from attending to more than the preliminary dressing of wounds, and all for whom transportation could be provided were sent forward to the Nashville hospitals. The ambulance trains, however, were not enough for all, and a good many had to be left in the field hospitals. For these, regular details of surgeons were assigned to remain in charge, with medical stores and supplies. On our return after the battle of Nashville, a fortnight later, most of our wounded thus left came again into our hands, together with great numbers of the Confederates who were not yet able to join Hood's army in its rough experience of a winter retreat beyond the Tennessee River.[1]

---

[1] Paper of Surgeon Frink, referred to on p. 183, *ante*, and published in Ohio Loyal Legion Papers, vol. iv. p. 418. Report of Medical Director, Army of the Cumberland, O. R., xlv. part i. p. 108. Reports of General Thomas, *Id.*, p. 41, and of Surgeon Heard for Fourth Corps, *Id.*, p. 174. An incident illustrating the pathetic scenes occurring in an army surgeon's experience was told me by my friend, Dr. Stephen C. Ayers, one of the surgical staff at the Cumberland Hospital in Nashville at the time of the battle of Franklin, and now Professor of Diseases of the Eye in the Cincinnati College of Medicine and Surgery. Among the wounded brought to the hospital from Franklin was a young soldier, a mere boy, who was unconscious from a gunshot wound in the head. The long ride from the battle-field had left him in extreme exhaustion, but Surgeon Ayers, noticing that he was still alive, attempted to give him some relief. He had been struck by a glancing ball which tore the scalp and left a well marked depression in the skull. After washing the parts and trimming away the hair, the surgeon, by the careful use of forceps, was able to lift the plate of bone and relieve the brain of pressure. The boy immediately opened his eyes, looking intelligently into the doctor's face. The latter seized the opportunity to ask his name and his home, and was answered clearly. In a moment, however, the clot pressed again upon the brain and consciousness was

The method of withdrawing our troops from the intrenched line was such as was usual with us. Skirmish and picket lines were under the charge of the Inspector Generals of corps and division, and a brigade officer of the day from each brigade reported with his detail for duty to the staff officer. Lieut. Colonel Hartsuff, General Schofield's Inspector General, was assigned to the general supervision of this duty, and Major Dow, my own inspector, met him at headquarters for consultation which should make thorough co-operation in the important task. The order of withdrawal was that which would put an end to my general command upon the line from the moment that the movement should begin; for no other order, written or oral, had modified or recalled those which had been issued in the morning, and which have been already quoted.[1]

Major Dow at once instructed the officers of the day on the Twenty-third Corps front as to their duty, and the whole skirmish line was carefully prepared for cool and intelligent obedience to orders. The Major himself was directed to collect and bring off the skirmishers after the last of the troops of the line had crossed the bridges.

The withdrawal of the artillery was not quite so simple a matter as marching off the infantry, for some batteries had suffered severely in horses, and refitting and rearrangement became necessary. Colonel Schofield, Chief of Artillery of the Army of the Ohio,

again lost. Before permanent surgical relief could be given, the young soldier died, but the moment of lucidity had saved him from a grave among the "unknown dead," and gave to his family the comfort of knowing the glorious end of his patriotic service and the means of identifying and reclaiming his body.

[1] *Ante*, p. 57, *et seq.* The effect of General Stanley's appearance at the Carter hill is considered in chap. xxi., *post*.

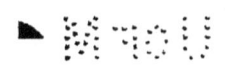

had been in frequent consultation with me during the day, and Captain Bridges, the Fourth Corps Chief of Artillery, had remained under my orders and with my staff at my position near the Carter house.[1] As it was desirable to get the batteries over the river before the infantry should leave the trenches, Captain Bridges was authorized to withdraw them one or two at a time, and as quietly as possible, beginning with those on the extreme flanks where there had been no serious fighting after dark.

Captain Scovill's Ohio battery, which was in reserve, and the regular battery under Lieutenant Canby, drew out at about eight o'clock.[2] Ziegler's Pennsylvania battery, and part of Baldwin's Ohio battery soon followed. Between nine and ten o'clock the Twentieth Ohio Battery, which had been in the focus of the fight back of the Carter house and garden, was withdrawn. Lieutenant Burdick, its commander, had been desperately wounded in the last effort of the enemy but a few minutes before, and the guns were under the command of Sergeant Horn. The loss in men and horses had been so severe that, at Captain Bridges's request, a detail of infantry men was furnished from Opdycke's brigade to haul the guns out to the turnpike by hand.[3]

Captain Marshall's Ohio battery next followed, but the Kentucky battery on the Columbia road under Captain Thomasson, and Captain Bridges's own Illinois battery under Lieutenant White (which was in position in Strickland's brigade), were still kept to strengthen the centre, and it was nearly midnight when the last of the artillery took its way to the bridge.[4]

[1] O. R., xlv. part i. p. 321.
[2] *Id.*, p. 338.
[3] *Id.*, p. 321.
[4] *Id.*, p. 327.

The actual movement of the infantry from the lines of intrenchment and over the river was a quiet one, and was not disturbed by the enemy. Some of the details, however, show how easy it is for mistakes to occur in orders which are issued without full knowledge. General Schofield's order necessarily gave only the outline of the movement, leaving minor matters to the judgment and discretion of his subordinates commanding the Fourth and Twenty-third Corps. It indicated the centre, on the Columbia Turnpike, as the dividing line, the troops on the right of it to cross the river by the foot-bridge and those on the left of it by the railway bridge.[1] No special mention was made of the artillery, but, in consultation with the chiefs of artillery, I arranged for the gradual withdrawal of this in advance of the infantry, as has been narrated.

Let us recall the actual position of the troops. On the right of the Columbia Turnpike, in the line, were Opdycke's brigade of Wagner's division, Ruger's division, and Kimball's division. On the left was my own division, now commanded by General Reilly. The two brigades of Wagner's division, with which he was in person, were at the river bank, and quite to the left of the Columbia road, and the central street of the village. They were in fact close to the head of the county bridge (the foot-bridge spoken of in the order), over which Wagner had placed a guard at my request. An earlier order, issued from Fourth Corps headquarters before the battle began, and when it was expected that the withdrawal would be at dark, had directed Wagner to cross by the railroad bridge and Kimball by the other.[2] The order for the midnight withdrawal had been issued

[1] O. R., xlv. part i. p. 1172.   [2] *Id.*, p. 1174.

from Fourth Corps headquarters in the form of a copy of General Schofield's, adding the further direction that "General Kimball will withdraw by his right flank, followed by General Wagner."[1] This sent them both by the county bridge, and assigned three divisions to cross there, and only one by the railroad.

Later in the evening the position of Wagner's two brigades in reserve seems to have been considered, and, as we learn from the diary of Colonel Fullerton, Wagner was ordered to take the advance in crossing. "December 1, 12.30 A.M. The troops of this corps commenced to withdraw from the line in front of Franklin, Wagner's division first, then Kimball's division, and to cross the Harpeth River."[2] Kimball does not appear to have been notified of this change, and on coming to the bridge he was surprised to find it occupied by Wagner's men. In his official report he says: "At midnight, in obedience to orders, I withdrew my division from its position, leaving my skirmishers on duty in front of the line, and moved to the bridge to effect a crossing, as I had been directed to move at once upon Brentwood to take up position till the army should arrive; but to my surprise I found the way blocked up by other troops who had left their position in advance of the time, and hence was unable to cross in advance, but was compelled to wait and take the position which others should have taken."[3]

From General Wagner's point of view, however, it was Kimball who was in the wrong, and in his official report he also refers to the circumstance, saying "About midnight, finding the troops which were to follow me across the river already crossing,

[1] O. R., xlv. part i. p. 1173.   [2] *Id.*, p. 151.   [3] *Id.*, p. 178.

I directed my brigade commanders to get ready and cross the river at once and march out on the Nashville pike."[1] He here refers to the two brigades with him at the river, for no orders were sent to Opdycke, as we shall presently see.

Kimball's supposition evidently was that the whole of Wagner's division had been in the line where Opdycke's brigade in fact was, and had that been the case he might well have been startled at finding the troops which were supposed to crown the arch which resisted the pressure of the enemy crossing the bridge before he who had constituted the most distant flank could reach it. The reader who has followed the narrative with care will not need to be told that no such compromising blunder had been committed. Wagner's two reorganized brigades in reserve were properly the troops first to leave the south side of the river; but Opdycke's brigade of the same division was still in place at the centre, between the two divisions of the Twenty-third Corps, grimly holding the line it had so glorious a part in restoring in the afternoon. And there it stayed till everything else from the right of the line had marched, and accompanied Ruger's division as it moved toward the bridge. "My withdrawal," says Colonel Opdycke in his official report,[2] "was under General Cox's instructions, and was accomplished at midnight." My own official report states the same fact. "The whole movement was made without interruption or molestation from the enemy, the third division (Reilly's) moving by the left flank and crossing the river upon the railroad bridge, and the second division (Ruger's) with Opdycke's brigade of the Fourth Corps, moving through the town, and

[1] O. R., xlv. part i. p. 232. [2] Id., p. 241.

crossing by a wagon bridge a little below the railroad crossing."[1]

From my headquarters at the Carter Hill,[2] I had kept track of the movement of the Fourth Corps divisions of which I have spoken. There was no guaranty that Hood might not again press in upon us if he suspected we were retiring, and it was a plain duty not to uncover the centre until Kimball should be out of the way, and the road to both bridges clear. As I have already explained,[3] the bridges and the ford were all near the left flank of our line, and Reilly's division alone could have drawn back through the town, covering all the crossings in case of need.[4] That division was therefore kept to the last in line, and after Ruger and Opdycke had marched into the town, Reilly's division faced to the left and moved through the trenches to the railway bridge. According to the rule in cases of retreat, I was with my staff at the rear of the column, and, except the pickets, we were the last to cross the river.

An incident had occurred late in the evening which made a vivid impression upon all who witnessed it. A fire broke out in the town, and a burning building lighted up the sky. Against this lurid background every man stood out a black figure more easily seen than in broad daylight. There could be no thought of moving till the fire should be put out. General Wood sent some of his staff from the north side of

[1] In my account of this movement given in "Franklin and Nashville," p. 94, I stated that the whole of the Twenty-third Corps crossed at the railway bridge. I was led to believe, by evidence based on the recollection of others, that this was the case, lapse of time having dimmed my own memory of the fact, and full access to official reports being then beyond my power. The fuller examination of the official records has corrected this error.

[2] O. R., xlv. part i. p. 355. See also Appendix B.
[3] *Ante*, p. 49. [4] See Map, p. 45, *ante*.

the river to superintend the efforts of the townspeople, and some detachments of soldiers. On the line the officers were busy keeping the men from exposing themselves and showing that we were on the alert, and it was a great relief when the dying light proved that the fire was not spreading, and that our movement would not be long delayed.

The fact that the enemy did not in any way disturb our withdrawal is sufficient evidence that none but the disabled and the dead remained near our works. Direct testimony, however, is not lacking. Major Dow tells of what he himself saw after the troops had marched off. He says: "When the command moved out of the works near midnight, I was left in charge of the skirmish line which was detailed to occupy them for another hour. I went over the works and walked some distance out in front. No enemies were there but those disabled or dead, and the cries of the wounded for help were very distressing. At that time I heard no signs of any force on either side of the pike. There had been occasional sharp volleys all the evening, but these had ceased, and there was no shot fired while our troops were withdrawn, nor during the time I stayed, or while I withdrew the skirmishers. We moved off entirely undisturbed, and overtook the command after daylight."[1]

General Schofield halted the troops as they crossed the river, and kept the whole in hand until all but the skirmish line were over. He then ordered me to take the advance with my division and march to Brentwood, the half way station to Nashville.[2] The orders of General Thomas implied that we should be

[1] See Appendix E.
[2] O. R., xlv. part i. p. 355, and Appendix B.

joined there by the reinforcements under General A. J. Smith.[1] We halted accordingly, and prepared to take position on the hills facing southward; but Smith's troops, disembarking from steamboats, were not supplied with wagon trains, and Steedman's, coming from Chattanooga, were delayed, so that General Thomas decided to make the rendezvous for the whole army within the lines of Nashville, and give some time to the increase of his cavalry.[2] The march was therefore resumed, and Nashville was reached by our advanced guard before noon. The rest of the troops arrived in good condition during the course of the day.

[1] O. R., xlv. part i. p. 1171. [2] *Id.*, part ii. p. 17.

CHAPTER XV

FRANKLIN AFTER THE BATTLE

Hood's Midnight Order — Condition of his Army — Discussion of Probabilities — Confederates move by the Flank — Experiences of the Carter Family — Colonel Carter's Story — Defensive Armor — Refuge in the Cellar — The terrible Night — Captain Carter's Fate — Private Gist's Adventures — General Cooper's Retreat.

HOOD'S stubborn purpose had not yielded to the terrible lesson of the day, and at midnight he issued orders for the renewal of the fight in the morning. He directed that his artillery be massed and open a concentrated cannonade upon our lines, and that his infantry be arrayed and ready at the signal to assault again at the point of the bayonet.[1]

It is not at all probable that his troops could have been led to a new assault if we had remained in our lines. The slaughter of the battle had produced a depression which is seen in every reference to it by those who were present. The itinerary of Cheatham's corps has a brief but very significant entry for the 1st of December: "To-day spent in burying the dead, caring for the wounded, and reorganizing the remains of our corps."[2] Brigadier General James A. Smith, who joined the army after the battle and succeeded to Cleburne's command,

---

[1] Hood's Report, O. R., xlv. part i. p. 654. Hood's Tennessee Campaign, by Major D. W. Sanders, Adjutant General, French's division, in Southern Bivouac for June, 1885, p. 13.

[2] O. R., xlv. part i. p. 731.

reported its great depletion in numbers, especially in officers, and added, "Nor was the tone and *morale* such as was desirable, owing to the fearful loss sustained in that battle."[1] Major-General Clayton's mention of "the terrible scourge which our brave companions had suffered" has already been referred to,[2] and is especially significant as coming from one whose own command had not been engaged. It indicates also his conviction of its uselessness, since he speaks of the merciful interposition of night which saved his men from a similar fate. Reading these things, one is convinced that Hood would have found his subordinates protesting against the renewal of what they plainly saw was a hopeless effort, and even if he persisted we should have repulsed him far more easily than on the previous day.

It must be remembered, too, that had we remained in our lines we should not have been idle. Our earthworks would have been strengthened. Reinforcements from Nashville would have been an essential part of our plan. A. J. Smith's corps, coming forward during the night, would have taken position on Hood's right flank, holding the east bank of the Harpeth, and enfilading with artillery the Confederate lines as they attempted to advance. At the first evidence of weakness or hesitation on the part of the enemy, these fresh troops would have been ready for a decisive flank attack, and we who were in the lines could have joined them in what might well have proved an anticipation of the crowning victory which was gained at Nashville a fortnight later.

Hood had no great reason to think he could have shelled us out of our position by a superior artillery

---

[1] O. R., xlv. part i. p. 739.   [2] *Ante*, p. 154.

fire in the early morning. He could not have had a more favorable opportunity than at Columbia a few days before, where my division had held him at bay, although his guns encircled the tongue of land which we held, and looked down upon us from the commanding ground around the bend of Duck River. Nerved by the knowledge that aggressive tactics on our side were being executed by strong reinforcements, we should not have been less tenacious of our position than we had been at Duck River. But the conclusion was not to be tried there, and we now know that the Confederate officers and men drew a breath of deep relief when the dawn of day showed that our trenches were empty.

Rendered prudent by his experience, Hood did not follow us by the direct turnpike to Nashville, thinking it quite probable that Schofield might turn and deliver a dangerous blow when part of the Confederate forces were across the Harpeth. He sent Forrest forward with the cavalry upon the Brentwood or Wilson Turnpike, and followed with Lee's and Stewart's corps, till the convergence of the roads made both available. One of his division commanders sarcastically reminded him that the same flank movement made on the day before would have saved the fearful losses which had crippled the army.[1] A skirmish between Forrest and Wilson was the only incident which enlivened the march to Nashville, and in a day or two Hood closed in upon the capital of Tennessee, impelled, we must believe, far more by a wish to give an air of victory to the engagement at Franklin than by any sound principles of military strategy. The heart of his army was broken, and

[1] The author was told this fact by a Confederate officer who heard the conversation.

Beauregard was unquestionably right (as was proved by the event) in saying, "It is clear to my mind that, after the great loss and waste of life at Franklin, the army was in no condition to make a successful attack on Nashville."[1] It was a sacrifice of military principle to the pride of the commander, and was naturally followed by a still greater disaster, which marked the beginning of the rapid downfall of the Confederacy.

The personal experience of the Carter family was a singular one, and illustrates so well the terrible straits into which private persons are thrust in time of war, that the narrative would hardly be complete without their story.[2]

I have already said that the head of the family was Mr. F. B. Carter, an aged gentleman,[3] with whom were living four grown daughters (one of them married) and a daughter in law. Of two sons, one, Captain Theodoric Carter, was on the staff of General Thomas B. Smith in Hood's army, and was in the centre of Bate's division as it advanced over his father's farm to attack our lines a little west of the homestead. The other son, Colonel M. B. Carter, was at home, a Confederate officer on parole, having been made prisoner in an earlier engagement. Three families of young grandchildren were also in the house, and a couple of female servants, making

---

[1] O. R., xlv. part i. p. 651.

[2] Besides my own very vivid remembrance of the anxiety of the family in the early part of the day of battle, and of their statements when we saw them again on our southward march a month later, Colonel M. B. Carter, the present proprietor of the place, has given me, at my request, a written narrative of what occurred in and about the house.

[3] *Ante*, p. 43.

a household of seventeen souls. To these were added a family of five, near neighbors, who sought the protection of the stout brick walls of the Carter house just before the combat opened. When the storm broke, therefore, there were three men and nineteen women and children, non-combatants, in the mansion.

Early in the day Mr. F. B. Carter had asked me, with some anxiety, whether he had better remove his family from the house and abandon it. Without knowing how large the family in fact was, I advised him not to leave the house unless it should become certain that a battle was imminent; for whilst my headquarters tents were in his door-yard, there was no danger of annoyance from the men of my command. If the house were abandoned, it would be impossible to answer for the safety of its contents. But if there were to be a battle, the very focus of it would certainly be there, and it would be no place for women and children. I thought it most probable at that time that Hood would not attack in front. The very thoroughness of our preparation to meet an assault was a reason why he should not make it. It seemed wise for the family to remain as they were till they saw that a battle was about to open, and then to hasten into the village.

Soon after our noon meal, my headquarters tents were struck and loaded on the wagon, which was sent to the rear. From this time I was in the saddle and saw nothing of Mr. Carter's household. His son, the prisoner on parole, naturally kept somewhat secluded, and avoided inquiries as to his own *status;* but when the peril was actually upon them, he became the leader of the terrified and helpless group. I shall let him tell the tale himself.

"While the preparations for the impending battle were going on," he says, "the Carter family were not inattentive observers. They had witnessed on other occasions sharp skirmishes between Rebel cavalry raiders and the Federal pickets stationed about the premises, in which men were killed and wounded, some in the yard, and even in the house itself. They felt themselves somewhat inured to the casualties of war; but the great number of men now so hurriedly and so intently engaged in demolishing houses and constructing works of defence, looked to them painfully ominous. The scene presented was on a bigger scale than anything they had ever seen before. It created feelings of profound anxiety. Whether to abandon home and the little that was left to them after three years and more of devastation, and to seek personal safety by flight, was the all-absorbing thought. In either aspect the prospect was discouraging. To leave home, pillage was almost certain, and blackened ruins might be all that would be left to greet their return. With one accord it was determined to remain: perhaps their presence would be respected and the house spared. They would trust to God to shield themselves from harm. . . .

"Although Hood was said to be a rash fighter, it was hardly thought he would be reckless enough to make a determined assault on the formidable works in front of him; but to be prepared for any emergency, it was directed that a bundle of clothing proportioned to the strength of each one be prepared, for the twofold purpose of having that much saved in case all else was lost, and for partial protection should they be forced to leave the house. If the latter became necessary, all were instructed to throw their respective bundles over their backs, and follow the leader whithersoever he led.[1]

"In a little while all doubts were solved as to Hood's intentions. His solid lines, to the right, to the left, and in front, advancing at a rapid pace, showed plainly enough that the crash was at hand. Although the house had withstood the shock of former conflicts, they seemed as child's play to the approaching storm. The cellar afforded the securest retreat, and hardly was it reached before the din of battle grew appalling. In the gloom of the cellar the children cowered at the feet of their parents, and while the bullets rained against the house, and a cannon ball went crashing through,[2] all seemed in a state of acute expectancy, but gave no audible sound of fear.

"The first onset having passed, and no one harmed, reassurance returned, and hope revived with some: with others, the comparative lull increased the tension and awakened fears of unknown dangers yet to come. In this state of alternating hope and fear, they dragged through the weary hours until the last shot was fired and deliverance assured. Mutual congratulations were scarcely exchanged when a Confederate soldier brought the sad tidings that Captain Theodoric Carter, a son and brother, lay wounded on the field. An elder brother, who had thus far directed affairs, went immediately in search, but by misdirection went to another part of the field.

[1] It was a shrewd device to provide such defensive armor if they had to run toward town with the hail of Confederate bullets coming from behind. Nothing would stop a bullet better than a well rolled bundle of clothing.

[2] The ball went through the wing or ell of the house, and the holes made by its entrance from the south and exit on the north are still there (1896).

In the mean time, General Thomas B. Smith, of whose staff young Carter was a member, reported the casualty and led the way, followed by the father, three sisters, and sister in law, to where the young officer lay, mortally wounded. They lifted him gently and bore him back to die in the home he had not seen for two years and more. He had fallen when his heart's wish was almost attained, only a few rods distant from the home of a lifetime."

When daylight came it was found that wounded men and some skulkers had possession of the principal floors of the house. Some of the wounded had crawled away from the lines in the darkness, and getting into the house lay down there and were overlooked by the stretcher bearers who were carrying the injured men to the rear. Others had been laid in order upon the ground to await the ambulances, already overloaded. These, of course, fell into the hands of the Confederates. The movement in retiring from the position at midnight was conducted so quietly that even some soldiers who were at the breastworks, overpowered by sleep after exhausting fatigue, did not get sufficiently awake to fall into line with their comrades. These were aroused and sent after the column by the pickets left to cover the movement.

Among the men of Lane's brigade, who turned and fought at our breastworks, was an intelligent young soldier, a private in the ranks of the 26th Ohio, who lived to become a college professor after the war, and to write a very clear and modest account of his personal recollections of the fight.[1] Mr. Gist was

[1] Professor W. W. Gist's paper, "The Battle of Franklin as seen by a Private," was published in the Cedar Rapids (Iowa) Weekly Republican, September 6, 1883. If all personal accounts of stirring

one of those weary ones who were overpowered by sleep, and we may accept his well told story of his experience as a lesson in the condition of an army which had been fighting and marching day and night.

"After the firing ceased," he says, "we waited for a long time in terrible suspense, expecting that the battle would be renewed at any moment. At last, wearied by the marching and the terrible scenes of the two days, and stupefied by the gunpowder, I fell asleep with my head resting against the works, I think. It must have been nearly midnight when I awoke. I was amazed to find that the line of battle was gone, and that only a skirmish line remained. I think that one of the skirmishers aroused me. At all events, he told me to hasten across the river, as the army was nearly all gone. I started without debating the question a moment. As I passed back of the line a horrible sight met my view. Here was a long line of wounded, lying as they had been placed, and moaning most piteously. They did not seem to understand that the army had gone, and that soon they would fall into the hands of the enemy. Reaching the pike, I started on the double-quick and soon crossed the river. The road was filled with soldiers moving toward Nashville, but there did not seem to be any organization. Almost as soon as I crossed the river, I met a man of my company, and to him I owe my escape. I was weary beyond anything that I had ever experienced, and hungry, and I could not walk a quarter of a mile without resting. As soon as I would sit down I would fall asleep, and in a few

---

events were equally candid and careful, the work of the historian would be a comparatively easy one. I shall have further occasion to refer to this valuable paper.

minutes he would wake me. This was repeated, I should think, a hundred times."

The physical powers of the soldier place a limit to the work which may be required of him. Excessive weariness is as absolute a barrier to vigorous exertion as would be mutilation by wounds or collapse from disease. It reaches, at last, a point where both mind and body become physically unreliable. Ordinary motives lose their power. The stimulus of ambition, of hope, even of duty, loses its power, and the body and mind imperatively demand rest at whatever cost. The weaker men and those upon whom some extra stress of exertion has fallen will show first the failure of powers, and straggling begins. Discipline will postpone mischievous results for a time, but not for long. It is hard to state any definite criterion of the endurance of men or horses, and historians sometimes fall into the error of assuming that marching and fighting may be kept up indefinitely. Mr. Gist's description of his own suffering is equally true of many more. Men have told me that on that night march they fell asleep as they walked, and found themselves stumbling or falling when a hollow in the road threw them out of poise. Continuous exertions for several successive days and nights in part of the command had brought them too near the limit of their strength, and this was a legitimate reason in favor of giving the troops a night's rest in the lines, even with the risk of further fighting on the morrow. It should be fully weighed in deciding whether it would have been wiser to push reinforcements forward to Franklin, rather than put upon exhausted men the additional strain of another night march in retreat.

The narrative of Schofield's army would not be complete without some further mention of the brigade under General Cooper, which we left in the vicinity of Centerville on the Duck River, some thirty miles west of Columbia, on the 29th.[1] Doubtful whether the messenger could get through who was sent by General Ruger to order Cooper to march to Franklin, Schofield requested General Thomas to send orders directly from Nashville to Cooper to the same effect. In the early afternoon of the 29th, therefore, Thomas sent a despatch by telegraph to Colonel C. R. Thompson, commanding the post at Johnsonville, on the Tennessee River,[2] ordering him to send a mounted messenger to Cooper at Centerville with an accompanying order to the latter to march by the Nashville road till he should cross the Harpeth at the farm of the Widow Demos, and from there report to Schofield at Franklin.[3]

Ruger's messenger reached Cooper at Beard's Ferry in the morning of the 30th, when Schofield's army was already assembling at Franklin, and Cooper marched the portion of his brigade with him to Killough's, five miles north of Duck River, sending orders to the rest of the command, which was at Centerville, to join him at once. The map seemed to make the distance between Centerville and Beard's

[1] Chap. ii., *ante*, pp. 24, 29.
[2] O. R., xlv. part i. p. 1163.
[3] In the official records the "Widow Dean" is the name given, but Cooper's report (*Id.*, p. 370) has the name "Demoss," which agrees with the name "Demos" on the map then used by the army, which had been compiled in 1862 by the Engineer Corps at Army of the Ohio and Cumberland headquarters. The crossing was where the Nashville and Hardin Turnpike reached the Harpeth. South of that river the road was a common country road to Centerville. The distance from Nashville to Centerville was supposed to be fifty-five miles, of which fifteen were north of the Harpeth.

Ferry some four or five miles; but Cooper found it fifteen by the shortest route,[1] and as the ferry was nearer to Columbia he had rightly established his headquarters there to shorten the time in communicating with Ruger and Schofield. The concentration of his whole command was not effected till noon of December 1st, when our little army was already at Nashville. At two o'clock Cooper received Thomas's message, which had been two days in reaching him. He immediately marched toward Nashville, and crossed the Harpeth at the place assigned on the evening of December 2d. During this day he learned from Rebel deserters that our army had fallen back from Franklin, and that his own route to Nashville was probably intercepted by Hood's army.[2]

At Franklin, on the 30th, Schofield learned from Thomas what orders had been sent to Cooper, and he sent a messenger down the Harpeth to order the latter to march at once to Nashville. But Cooper, of course, was still near Centerville, and the messenger was chased by Confederate patrols into Nashville.[3] Upon this, Thomas telegraphed to Lieut. Colonel Sellon, commanding the post at Kingston Springs on the railway leading to Johnsonville, to send a mounted party to Centerville at once, to direct Cooper to make for Clarksville on the Cumberland River forty miles west of Nashville, which was made the rendezvous for the posts along the Tennessee which might be cut off from Nashville by Hood's advance.[4] This message did not reach

[1] O. R., xlv. part i. p. 370.   [2] *Id.*, p. 371.
[3] *Id.*, pp. 1171, 1175. In the dispatches between Thomas and Schofield the name of the crossing is again given as the "Widow Dean's," and Schofield says he can find no such place on the map. It was probably misspelled in telegraphing.
[4] *Id.*, pp. 1194, 1195.

Cooper, who continued his march, as we have seen, reaching the Harpeth in the evening of December 2d. He now marched cautiously toward Nashville on the Hardin Turnpike, until he reached a high hill eight miles from the city, from the top of which he saw the numerous camp fires on both sides of the road in front of him. He then moved across country to his left and rear till he reached the Charlotte Turnpike, which follows the general westerly course of the Cumberland River, and, marching hard upon it all night, he recrossed the Harpeth at daybreak of the 3d. His own good judgment now indicated Clarksville as his objective point, and he continued his movement in that direction. He arrived there on the evening of the 5th, joining Colonel Thompson's Johnsonville garrison, which had just preceded him under orders from General Thomas. Resting two days, Cooper resumed his march to Nashville, where he rejoined the Twenty-third Corps, and reported to General Schofield on the evening of the 8th. Instead of fifty-five miles, his devious march from Centerville had stretched to two hundred and ten, and might easily have ended in captivity; but his coolness and good judgment had made a successful escape from the perils and difficulties of his situation, and good discipline, with the good marching qualities of his veterans, had brought the command in without loss.[1]

[1] Cooper's Report, O. R., xlv. part i. pp. 370, 371.

CHAPTER XVI

RESULTS AND LESSONS

Sources of Statistical Knowledge — Hood's Forces before the Battle — Schofield's — Numbers actually engaged — Hood's Casualties — Loss of Officers — Schofield's Losses — Analysis of them — The Problem of Attack and Defence — Fire Discipline.

It is not easy to be very exact in stating the numbers of men engaged at Franklin in the two armies under Hood and Schofield, or the numbers of casualties in the battle. On the side of the Confederates there was a natural disposition to avoid too great explicitness; for the results of the battle produced a sensible shock throughout the Southern States, and it was generally looked upon as a fatal blow to their hopes of independence. On the national side the army under Schofield did not have a permanent organization, but was, in some sense, an accidental assemblage of such portions of the troops under Thomas in the Military Division of the Mississippi as were at the time in hand for operations against Hood's invading army. By carefully comparing the reports of strength now published in the Official Records, we can, however, eliminate most of the sources of error, and make approximations that are very close to accuracy.

The tabulated reports of Hood's army on which we must chiefly rely are those of November 6th and

December 10th.[1] No intermediate reports seem to have been preserved. There had been a little desultory fighting on the way northward from the Tennessee River to Columbia, and sharp affairs at the crossing of Duck River and at Spring Hill. It is probable, however, that the casualties prior to the battle of Franklin were more than offset by recruiting; for one of the avowed advantages which Hood expected to gain in Tennessee was the increase of his army both by voluntary enlistments and by the enforcement of the Confederate conscription laws.[2] There is evidence that up to the battle of Franklin, at least, there was considerable success in gathering up men who had straggled homeward from the Tennessee regiments, and in bringing in fresh men who were liable under their laws to military duty.[3] We shall therefore be probably within the mark if we assume that the difference between Hood's morning reports of the dates above given is due to the casualties of the battle of Franklin, and check this by the reports of Generals Thomas and Schofield of the direct information obtained at Franklin on our return there after the battle of Nashville.

On the 6th of November Hood reported his infantry and artillery present for duty to be 32,861.[4] Jackson's division of cavalry was 2,801. But Forrest's cavalry corps consisted of three divisions, the commanders being Jackson, Buford, and Chalmers; and there is no doubt that Buford's division was a larger one than Jackson's. It is most probable that the average strength of Forrest's divisions was 3,000,

[1] O. R., xlv. part i. pp. 678, 679.
[2] Hood's "Advance and Retreat," p. 305.
[3] Dispatch of Colonel Blake to General Rousseau, O. R., xlv. part i. p. 896.
[4] Id., pp. 678, 679.

making the strength of his corps 9,000. Notes upon Hood's original returns show that Forrest made no tabulated reports of his corps through the whole campaign, and we have to rely on approximate calculations. Gathering the best information that could be got from prisoners and deserters, and comparing it with other evidence, General Schofield stated in his own official report that the smallest estimate of Forrest's corps was 10,000, and this he thought was nearly accurate.[1] The total strength of Hood's army, then, was from 42,000 to 43,000 of all arms.

The data for determining the strength of General Schofield's army are more complete, since we have the regular tri-mouthly morning reports. As to infantry and artillery, these would be all we could wish if the published tables gave in detail the strength of detachments, such as that under General Cooper, who marched from Centerville on the 30th of November, with his brigade and two additional regiments belonging to the Twenty-third Corps.[2] As Cooper did not rejoin the corps until it had reached Nashville, his command must not be counted as among the combatants at Franklin. No return of the Union cavalry with General Schofield was separately made till the morning of the 30th November, and even then it was only provisionally organized, so that it is difficult to separate it from other parts of the cavalry corps at Nashville and elsewhere. The discrepancies as to the cavalry on either side are, however, of minor importance, as the strongly dominant feature of the battle of Franklin was the infantry engagement.

On the morning of November 20th, the infantry and artillery of the Fourth Army Corps, present for

[1] O. R., xlv. part i. p. 341. See Appendix A.   [2] *Id.*, p. 367.

duty equipped, numbered 13,119; of the Twenty-third Army Corps, 9,823: total, 22,942. On the morning of the battle, November 30th, the numbers were, of the Fourth Corps 15,207, of the Twenty-third Corps 10,527, besides two regiments of the Army of the Tennessee (44th Missouri and 72d Illinois), detached and temporarily assigned to the latter corps.[1] These two regiments were more full than most of the old organizations, and numbered, perhaps, a thousand men.[2] The gain of 2,792 in the aggregate of the two corps consisted of new regiments and recruits which joined Schofield at Columbia. But the absence of Cooper's command reduces the number present at Franklin in the Twenty-third Corps by about 3,000 men, and makes the aggregate of Schofield's infantry and artillery at the opening of the battle 23,734.

The national cavalry under General Wilson must be estimated in part, for reasons already given. Hatch's division is reported as numbering 2,638, Johnson's 2,286, and Hammond's brigade is not reported.[3] The aggregate was about 5,500, and the total strength in all arms of Schofield's army 29,234.

For the purposes of military criticism and information it will be profitable to consider also the forces on both sides actually employed in attack and defence. Of Hood's army two divisions of Lee's corps were held in reserve and not engaged, and one brigade of Stewart's corps (Ector's) was absent, escorting the pontoon train. Only two batteries of his artillery appear to have been under fire. The numbers of Hood's infantry and artillery actually

[1] See *ante*, p. 54, and Ruger's Report, O. R., xlv. part i. p. 367.

[2] On the concentration of the army at Nashville, these two regiments were transferred to General A. J. Smith's "Detachment of the Army of the Tennessee."

[3] O. R., xlv. part i. p. 54.

delivering the assault on our lines were, therefore, two or three hundred less than 24,000.

The assault was received upon the continuous intrenched line between the Harpeth River above the town on the left, and the Carter's Creek Turnpike on the right. As we have seen,[1] Kimball's division of the Fourth Corps was rather a guard or cover for the right flank than a part of the continuously intrenched front, and, except part of its left brigade, it was only slightly engaged with Chalmers's division of the enemy's cavalry. It, as well as our cavalry engagement across the river on the east, might properly be dropped from the consideration of the conditions and results of the infantry assault by the Confederates, though both must of course be considered in the statistical results of the whole engagement. I shall analyze the latter first.

Upon our reoccupation of Franklin, December 18th, unusually full opportunities for knowledge of the enemy's losses were made available. Their hospitals fell into our hands, and we recovered our own, with our medical officers and attendants who had been left in charge of them. General Schofield was thus enabled to report definitely that Hood's losses were "1,750 buried upon the field, 3,800 disabled and placed in hospitals in Franklin, and 702 prisoners, making 6,252 of the enemy placed *hors de combat*, besides the slightly wounded. The enemy's loss in general officers was very great, being six killed, six wounded, and one captured."[2] As to

---

[1] *Ante,* p. 60.

[2] O. R., xlv. part i. p. 344, and Appendix A. General Thomas's report gives the same figures. *Id.,* p. 35. My own investigation on the field at the time made the killed 1,800, the wounded in hospital, 3,800, and the number of prisoners captured by the Twenty-third Corps,

the slightly wounded, Hood's own dispatches to the Confederate War Department are conclusive. On the 5th of December, trying to remove the stunning effect of the terrible loss in general officers, he telegraphed to the Secretary, Mr. Seddon: "Our loss of officers in the battle of Franklin on the 30th was excessively large in proportion to the loss of men. The medical director reports a very large proportion of slightly wounded men."[1] Any accepted ratio between killed and wounded which military experience has established, will show that the temporarily disabled who had been treated in their quarters or who had left the hospitals and returned to the ranks before Hood's next return of his force was made (December 10th) would add about two thousand to the list of casualties and swell the aggregate to eight thousand or more.

It is interesting, in confirmation of this, to compare the Confederate returns of December 10th with those of November 6th, which have been analyzed above. The returns of infantry and artillery of the later date show 24,074 present for duty.[2] To these must be added four absent brigades, which at the proportionate part of the divisions to which they

---

702. I added: "Thus, without estimating the prisoners taken by any part of the Fourth Corps, or the stragglers and deserters, who are known to have been numerous, the enemy's loss was not less than 6,300." Official Report, *Id.*, p. 356, and Appendix B. General Wagner's Report (*Id.*, p. 232) says his brigade commanders claim 753 prisoners. In his special report of prisoners, flags, etc., he adds the postscript: "The above is the report of brigade commanders. Only about 400 prisoners (officers and men) passed through the hands of my provost marshal." *Id.*, p. 234. The reports of Generals Schofield and Thomas include only the prisoners reported by me for the Twenty-third Corps. See also chap. xix., *post*.

[1] O. R., xlv. part ii. p. 650.
[2] *Id.*, part i. p. 679.

belong would amount to 2,387,[1] making the aggregate 26,461.[2] The difference between this and 32,861, the similar aggregate for November 6th, is 6,400, which would show the losses in infantry and artillery. To these must of course be added those of the cavalry, which are reported as being 269 for the month of November. The total is thus raised to 6,669, which excludes all cases of slight wounds where the men had returned to duty within ten days after the battle.[3] The resemblance of these figures to those of General Schofield's report made at the time is very striking.

It would, of course, be very desirable to have full official returns of the Confederate losses, and thus be saved the need of comparison and calculation; but no reports are preserved, if ever made, of Cheatham's divisions (Cleburne's and Brown's), which suffered most heavily, and we are therefore obliged to have recourse to the kind of evidence which has been presented.

The thing which most impressed the world at the time was the unusual number of general officers who appeared in the casualty lists. Hood gave the list with laconic brevity in his dispatch to the Confederate War Department on December 3d. "We have to lament," he said, "the loss of many gallant officers and brave men. Major General Cleburne, Brigadier

[1] A striking example of the destruction of some of the brigades is that of Cockrell's of French's division, one of those included in the absentees on December 10. It went into the fight with 696 officers and men, and suffered 419 casualties. O. R., xlv. part i. p. 716.

[2] *Ante*, p. 211.

[3] There are discrepancies in the reports of the cavalry which cannot easily be reconciled. The report quoted gives the losses of Chalmers's division for the month of November as 35 (*Id.*, p. 762); but Chalmers's report makes them 116 killed and wounded in the battle of Franklin. *Id.*, p. 764.

Generals John Adams, Gist, Strahl, and Granbury were killed: Major General John C. Brown, Brigadier Generals Carter, Manigault, Quarles, Cockrell, and Scott were wounded; Brigadier General Gordon was captured.[1]

But this was only a part of the response to his order immediately after the battle, that "Corps commanders will send in at once a list of the division, brigade, and regimental commanders by name and rank, who were killed or wounded so as to be unfit for service, in the engagement of yesterday evening."[2]

The complete return is a roll of honor which fills nearly three pages of the published official records, and of which the summary is five general officers killed, six wounded, and one captured; six colonels killed, fifteen wounded, and two missing; two lieutenant colonels killed and nine wounded; three majors killed, five wounded, and two missing; two captains killed, three wounded, and four missing: a total of sixty-five.[3] Remember that none of these were exercising a less command than that of a regiment. Every captain on the list was serving three grades above his rank when he fell, stepping forward to fill vacancies as they had been made by the fortunes of war. The other field and line officers who fell are mingled in the long list of the thousands in which the sacrifices of war are summed up. When silence fell upon the field, there was more than one brigade in which a captain was the ranking officer in command.[4]

---

[1] O. R., xlv. part ii. pp. 643, 644.
[2] *Id.*, p. 629. [3] *Id.*, part i. pp. 684-686.
[4] This was the case in Quarles's brigade of Walthall's division, and in Gist's brigade of Cheatham's division. *Id.*, pp. 721, 738.

The losses on the national side are more readily ascertained from the official reports, for the complete aggregate by divisions in the Fourth and Twenty-third Corps was reported by General Schofield.[1] Those of the cavalry are returned for the whole campaign, including the battle of Nashville, so that the losses of General Wilson's command at Franklin cannot be separately stated.[2] It was understood that they were not large. General Schofield's tabulated statement is as follows: —

	Killed.	Wounded.	Missing.	Aggregate.
Fourth Army Corps:				
First Division (Kimball) .	5	37	18	60
Second Division (Wagner)	52	519	670	1,241
Artillery	10	51	6	67
Twenty-third Army Corps:				
Second Division (Ruger) .	74	241	313	628
Third Division (Reilly) .	48	185	97	330
Total	189	1,033	1,104	2,326

The analysis of this total of 2,326 casualties shows several important and instructive facts.

First, nearly half of the whole are in the list of "missing." Whilst it is true that among the missing would be some of the dead who fell outside the works, and a few stragglers and fugitives from the battle-field who did not rejoin their colors, the greater part would be men, wounded or unwounded, who fell into the enemy's hands. In his dispatch of December 3d, Hood claims that he captured about a thousand prisoners. His claim is corroborated by this list of 1,104 missing.[3]

Second, nearly the whole of the missing are in Wagner's and Ruger's divisions, and nearly two

---
[1] O. R., xlv. part i. p. 343.   [2] *Id.*, pp. 47, 568.
[3] *Id.*, part ii. p. 643.

thirds of the whole are in Wagner's alone. Analyzing still further by brigades, the number of missing in Moore's brigade of Ruger's division is trifling; so is it in Casement's and Stiles's brigades of Reilly's division.[1] As to the Twenty-third Corps, therefore, the missing are nearly all from Reilly's and Strickland's brigades, the two which were respectively on the right and left of the Columbia Turnpike, and there were three times as many in Strickland's as in Reilly's brigade. In Wagner's division Opdycke's brigade returns 70 missing,[2] and the 600 remaining are from Lane's and Conrad's brigades. Those in Opdycke's brigade are almost exactly the same as in Reilly's brigade. These figures demonstrate that where the line was unshaken and the troops held firmly to the works, the loss was small; where the break was momentary and the line quickly restored (as by Reilly and Opdycke), the prisoners lost were still moderate in number; where the break was more complete and the enemy held the works for a longer time (as in Strickland's line), the losses were large; but where the enemy came in pell-mell with our men who were driven in from the position nearly half a mile in front, both wounded and missing were multiplied, as one would naturally expect.

Third, the lists of the killed tell a similar story. They are trifling everywhere but near the centre, on the national side, though the Confederates lost nearly as heavily in front of Stiles and Casement as anywhere on the field.[3] It helps to show the character of the fighting in Strickland's brigade when we find that the number of the killed in it was 53, whilst

---

[1] O. R., part i. pp. 381, 409.   [2] *Id.*, p. 241.
[3] See *ante*, pp. 124, 126.

the number killed in Lane's and Conrad's brigades together was 36.[1] In Opdycke's brigade 16 were killed, making 52 in the three brigades of Wagner's division to 74 in the two brigades of Ruger's. This shows not only that Strickland's regiments rallied stoutly and fought stubbornly, but that the second hastily constructed breastworks were poorer cover for them than the original works in the main line. Thus every column of the tables corroborates the narrative of the battle as we have read it in the preceding pages.

The improvements in repeating arms made since our civil war, and the current discussion of the practical range and rapidity of fire from a line of battle, receive light from our experience at Franklin. We found that the slight undulations of the field were scarcely noticeable from our parapet, and yet they were sufficient to cover Hood's advancing lines of infantry so well that it was not till they had passed the position first occupied by Wagner's two brigades that they came under infantry fire. They thus got within five or six hundred yards of our lines practically unharmed from musketry. We rarely found a field, during the war, so open or so level as this, and one might fairly be sceptical as to the practical value of much greater range in small arms.

As to rapidity of fire, however, the proof seems strongly in its favor. The few repeating rifles we had bore no important ratio to the number of men in line, though the enemy, exaggerating the number of such weapons, credited them with much of the terrible destruction of the field. The truth was that the crowding of our second line and reserves into the

[1] O. R., xlv. part i. p. 381, and table, *ante*, p. 215.

works practically made all our weapons repeaters. For as the men were three or four deep in most places, they supplied the front rank so rapidly with loaded pieces that I doubt if any ordinary line armed with the latest magazine gun could have delivered so continuous a fire as we witnessed. As darkness came on, the appearance was so exactly that of a sheet of fire lying stationary and uninterrupted at the level of the parapet, that the engagement is rarely mentioned by one who was there without speaking of this, a striking phenomenon of the battle. With the weapons of to-day a similar result would be produced by a line in two ranks.

The engagement also sheds instructive light on the question, much mooted in recent military discussions, of the limit of endurance of loss by well disciplined troops, beyond which they cannot go. No one competent to judge denies that for a union of personal courage with intelligence, the material of the Union and Confederate armies in 1864 cannot be excelled. Their fire discipline was also of a very high order. Something less than 24,000 of such men[1] actually delivered the Confederate assault upon the line of breastworks between the Harpeth River and the Carter's Creek Turnpike, say a mile long, held by about 10,000 national troops, made up of Reilly's and Ruger's divisions and the artillery.[2] About 5,000 more (Wagner's division) were in the fight; but the preceding narrative makes it clear that, under the circumstances, their participation was a misfortune and not a benefit to the national army. The incidents at the opening of the battle greatly increased the confidence of the Confederates, and stimulated their hopeful courage far beyond

[1] *Ante,* p. 211. [2] *Ante,* p. 210.

the ordinary. Their total casualties, including the slighter but temporarily disabling wounds, were about one third of their whole number before they yielded to the inevitable and ceased their efforts. It would be hard to find a better test of what courage, nerve, and discipline are capable of.[1] It helps to establish a practical limit where the sense of hopelessness and impossibility quenches will. Beyond that point continued struggle is not heroism, it is insanity. The men who held the works and restored the line when broken, may also fairly claim that they too showed what nerve and will may do to retrieve an error and turn a threatened disaster into victory.

Among the trophies of the battle General Schofield was able to report thirty-three flags. Twenty of these were captured along Reilly's parapet, ten at Opdycke's, and three at other points. Many of these were taken in hand to hand conflict, but some were dropped upon the field, where the fearful destruction left no one to lift them again. Hood also claimed several, taken in the rout of Wagner's outpost before the rally at the main line renewed the continuous front of fire and steel.[2]

[1] An interesting discussion of the general subject is found in the papers of Captain F. N. Maude, R. E., republished by Captain Arthur L. Wagner, U. S. A., in the "International Series," vol. i.
[2] See Schofield's Report, Appendix A, pp. 306, 307. Opdycke's Report, O. R., xlv. part i. p. 242. Hood's Report, *id.*, p. 658.

CHAPTER XVII

DISCUSSION OF WAGNER'S CONDUCT

Natural Rise of Controversies — Corps Feeling — Good Comradeship — Wagner's Personal Situation — Disposition to befriend him — Criticisms by his Subordinates — Efforts to allay the Irritation — Correspondence — Preliminary Reports — Conrad's Report — Wagner retired from the Division — Leaves the Army.

I HAVE now completed the connected narrative of the battle of Franklin with what may seem abundant fulness of detail; but, as I have had to note on several occasions, controversies have arisen in regard to various points in the history which may be conclusively settled, as it seems to me, by a careful weighing of the evidence, official and private, which we now have access to. One of the most interesting features of a civil war waged by a people of so general intelligence and mental alertness as ours is the great activity of discussion which follows it. Our old soldiers were keen critics of what they saw, and in the long delay in publishing the official records of the war debated its issues and fought its battles over with none the less zest because their information was limited and their memory imperfect.

As it fell to my lot to publish in 1882[1] a comparatively brief history of the campaign of which the battle of Franklin was a part, it was an altogether natural result that I should find some of my state-

[1] *Ante*, p. 2, note.

ments and conclusions challenged. The brevity of my narrative made it impossible to give the evidence which supported it. I was obliged to state facts and conclusions succinctly, trusting to a confident belief that candid investigation would show that I had asserted nothing without strong proof. My personal familiarity with the events of which I wrote gave me some advantage in following the clues which lead to the truth. On many points I was also an eyewitness who owed it to himself to be explicit in telling what he knew, leaving it to the reader to give proper weight to his testimony.

I have little reason to complain of my treatment as an author. The most competent judges have accepted the authority of my volumes, and the publication of the government records has practically quieted discussion. There are, however, several subjects still debated which should be settled, if possible; and those who are interested in historical investigation will, I hope, be glad of the additional light I may be able to give.

A very natural *esprit de corps* led some worthy men who served in Wagner's division to question the justice of the criticism upon his handling of the division in the battle. The quotation from the official records of the written orders given him and his own report of his understanding of them fully disposes of the contention that the responsibility for his giving battle in front of our lines of works should rest upon his superiors. There have, however, been occasional allegations of a disposition on the part of General Schofield and other officers of the Twenty-third corps to be unfriendly to him in the examination which General Thomas instituted on the arrival of the army at Nashville. The fact was quite the opposite. General Schofield was well aware of jealousies which grew out of his assignment to the command of the army in the field, and had sought by every means in his power to allay them. In this he was heartily seconded by his principal subordinates in his own corps. General Ruger had just been transferred from the Army of the Cumberland to the Twenty-third Corps, and could not be otherwise than warm in his relations to his old comrades. My own had been scarcely less so from the day we had welcomed Wood's and Sheridan's divisions as reinforcements near Knoxville in the winter campaign of 1863-64. How we proved this cordiality will be shown in the events of the week following the battle.

It was past noon on the 1st of December when we reached Nashville, and the troops were hardly put in position about the town when General Wagner called upon me. He was in distress over the results of the engagement to his division, and was most anxious to soften the judgment of his superiors as to his action under his orders. His statement regarding the orders he gave to his own subordinates was that which he embodied in his official report the next day.[1] It was, in brief, that they were directed "to develop the enemy, but not to attempt to fight if threatened by the enemy in too strong a force." But the enemy's "onset was so sudden that Colonel Conrad found his brigade heavily engaged and about to be flanked before he could withdraw his line. His men fought gallantly, refusing to retire till completely flanked and driven out of their hastily thrown-up barricades, when they retired on the main

[1] See quotations in chap. iv., *ante*, and the whole report, O. R., xlv. part i. p. 231.

line." [1] There was no room for disagreement between us, and there was none, as to the rallying and reorganization of Conrad's and Lane's brigades near the river, or as to Opdycke's charging to restore the lines at the centre. General Wagner's anxiety was to get me to accept the view that his two brigades made a retreat, in accordance, substantially, with his orders, and took a position within the lines where they would be in reserve supporting my line in the works.

Wagner had seen a good deal of active field service, and under a rough exterior had a generous and genial character. If his account of the matter were accurate, it was rather the excess of courage in the two brigades which had compromised them, and there was no occasion for blaming him. To one who was friendly, it was easy to yield to his appeal for help in setting his conduct in a favorable light. There had been no time to investigate. We had not even rested from the fatigue and the excitement of the past three days. I had seen the first break of his men from the outer line, but I had not seen and had not yet learned the particulars of what occurred whilst I was riding from the left to the centre. I felt at liberty, therefore, to accept his view, and to promise him the assistance of a friendly report, made provisionally and expressly preliminary to an official one which should come later, when subordinate reports should be received and investigation be completed. General Schofield accepted the same favorable views; and the preliminary reports which went to General Thomas immediately, attributed the momentary break at the centre to the cover afforded the enemy by our retreating outpost, and the mis-

[1] O. R., xlv. part i. p. 231.

understanding of orders by the troops in the breastworks near the turnpike, in the confusion when Wagner's two brigades passed over them.

I should hesitate to state what occurred at this visit of an officer now dead, were there not contemporaneous written evidence of it, so that it does not depend on my recollection.[1] The substance of his conversational statement I have been content to put in the language of his own official report; though it was more earnestly emphatic in the freedom of such an oral interview, and claimed that his two brigades made a leisurely retreat. He was well aware of severe criticism among his subordinates, and that it was likely to lead to official inquiry if not to a court-martial. This was in fact the motive for his so great haste in seeking to remove unfavorable impressions from the minds of those immediately above him. Feeling that the glorious result of the engagement made it easy to take the favorable view of his action, as he stated it, and to treat it rather as an accidental mishap than as a disobedience of orders, I was entirely cordial in giving him prompt friendly assistance and counsel.

The tart exchange of opinions between himself and Colonel Opdycke at the time the brigade was marched within the lines,[2] I advised him to ignore; and as I was personally witness to Opdycke's heroism in the magnificent charge of his brigade, I urged him to do full justice to it, and to recommend the colonel's promotion. As he had been busy with the reorganization of the other brigades, and had not personally seen Opdycke's conduct, I agreed to take the initiative and write the recommendation for promotion which he would sustain in his report. With

[1] *Post*, p. 226.   [2] *Ante*, p. 73, note.

regard to the irritation in Conrad's and Lane's brigades, I comforted him with saying that, the facts being as he stated them, a full recognition of the courage which had been only too persistent would be likely to allay the irritation and bring the whole into harmony again. I did not then know the incident which my Adjutant General had witnessed at Franklin, when General Wagner, haranguing his partially reorganized troops near the river, had said, "Stand by me, boys, and I'll stand by you." The effort to overcome the insubordinate feeling began there.[1]

The contemporaneous documents to which I have referred are so necessary a part of the narrative that I give them in full. First is my official recommendation of the promotion of Colonel Opdycke.

"HEADQUARTERS THIRD DIVISION, TWENTY-THIRD ARMY CORPS, NASHVILLE, TENN., Dec. 3d, 1864.

"MAJOR GENERAL THOMAS,
  Commanding Army of the Cumberland.

"GENERAL, —

"I have the honor of calling to your notice the conspicuous gallantry and meritorious conduct of Colonel Emerson Opdycke, commanding a brigade in General Wagner's division, Fourth Corps, in the battle of the 30th ultimo.

"In the early part of the engagement a portion of the Second Division, Twenty-third Corps, was somewhat disordered by misunderstanding the purpose of retiring through the lines of the two brigades of Wagner's division which had been engaged in advance. It was at this time that Opdycke's brigade was lying in reserve on the Columbia pike, and, being called upon, he led them forward in the most brilliant manner, charging the ad-

---

[1] *Ante*, p. 146, and Appendix F.

vancing rebels and restoring the continuity of our lines, which were not again disordered in the slightest degree. Colonel Opdycke's promptness and courage deserve official notice, and, as I was commanding the line at that time, I deem it my duty to bring the facts to your attention, the more especially as he does not belong to my division.

"Very respectfully, your obedient servant,
"J. D. Cox,
Brigadier General Commanding."

This letter was sent, by the usual military rule, through the headquarters of General Schofield, and is still on the files of the War Department.[1] The recommendation was adopted and indorsed by General Schofield in his own report, dated December 7th.[2]

I caused a copy of this to be made and sent it to General Wagner with the following letter of my own: —

"HEADQUARTERS THIRD DIVISION, TWENTY-THIRD ARMY
CORPS, NASHVILLE, TENN., Dec. 3d, 1864.

"GENERAL, —

"I have the honor of handing you herewith a copy of a paper forwarded by me to General Thomas's headquarters, concerning the subject matter of which I had a conversation with you a couple of days since. I think this would be a proper time to press the subject of Colonel Opdycke's promotion, and if this can be made the means of doing so, I shall be most happy. Please inform me what steps you think can be advantageously taken, and it will give me great pleasure to co-operate further.

"I desire also to express my admiration of the gallantry of your whole command. Indeed an excess of bravery kept the two brigades a little too long in front,

[1] O. R., xlv. part i. p. 409.
[2] *Id.*, p. 343, and see Appendix A.

so that the troops at the main line could not get to firing upon the advancing enemy till they were uncomfortably near.

"Very truly and respectfully,
Your obedient servant,
"J. D. Cox,
Brigadier General Commanding."

"To BRIGADIER GENERAL G. D. WAGNER,
Commanding 2d Division, Fourth Army Corps."

This letter, which by its tenor General Wagner was permitted to use for the friendly purposes already stated, was not strictly official, and does not appear to have been preserved in the Official Army Records or in my private letter-book. Finding an extract from it quoted in one of the many newspaper articles upon the battle, Major E. C. Dawes (the well known military statistician and writer) procured a copy attested by an officer of General Wagner's staff. It fixes the date of the interview between General Wagner and myself on the 1st of December, the day after the battle, when, after our twenty-mile march begun at midnight, we went into position in the fortifications about Nashville in the afternoon.

In connecting the two papers now given, I have outrun the order of events and must return to the 2d of December. On the morning of that day General Schofield informed me that General Thomas desired a preliminary sketch of the battle for immediate use, and, although it was quite impossible to get subordinates' reports, asked me to write it at once, as the officer who had fullest and most direct personal knowledge of what occurred upon the line. I did so without a moment's delay, and it was forwarded to General Thomas, bearing that date.[1] General

[1] O. R., xlv. part i. p. 348. See also Appendix A.

Schofield made a report for the army, also preliminary, on the 7th of December; but neither of the regular official reports from the corps headquarters was made for more than a month afterward.[1]

In this preliminary report, thus made without time for investigation or official data, and both expressly and impliedly subject to full correction in the official one which was to follow, I felt not only at liberty to follow my friendly feeling toward General Wagner, but, with only his statements of the preceding evening before me, I felt it due to him to accept them in regard to the retreat of his two brigades. In this clause of the report, therefore, I used his freshly uttered words, and said, "At three o'clock the enemy engaged the two brigades of Wagner's division, which, in accordance with orders, fell leisurely back within our lines, and the action became general along the entire front."[2]

Everything, therefore, which kindly personal feeling or sympathetic comradeship could do to assist General Wagner had been done in the Twenty-third Corps. How did it happen that it did not avail?

On the evening of the 1st of December,[3] before resting from the fatigues of the battle and the march, Colonel Conrad made his report for the Third Brigade of Wagner's division. He had not waited for the reports of his regimental commanders, the earliest of which are dated on the 5th.[4] Of the other brigade reports, Opdycke's was dated on the 5th and Lane's on the 7th.[5] As I have already said, this haste was

[1] The regular official report of the Fourth Corps was made by General Thomas J. Wood, from Huntsville, Ala., on the 10th of January, 1865, and mine for the Twenty-third Corps was made from Clifton, Tenn., on the same date. O. R., xlv. part i. pp. 119 and 349.
[2] O. R., xlv. part i. p. 349.
[3] Id., p. 269.
[4] Id., p. 275, etc.
[5] Id., pp. 239, 255.

not only unusual, but it was equivalent to preferring charges against the division commander [1] when the character of the report is considered. It anticipated and contradicted General Wagner's theory that the two brigades had been compromised by the persistence of its officers and men in fighting when their orders directed their withdrawal, and asserted that the general's orders had been explicit, even to directing the file-closers to hold the men to the line with fixed bayonets.[2] That this was intended to be followed up by personal complaints at General Thomas's headquarters needs no argument. General Wagner evidently so understood it, for he hastened his own report and presented it on the next day, three days before any other report from his subordinates was received. There is internal evidence that he sought in it to soften the collision between his own and Colonel Conrad's statements, but the consequences speak for themselves.

In Colonel Fullerton's Itinerary of the Fourth Corps, under date of 3d December, this entry is found: "Brigadier General Elliott has been assigned to the Second Division of this corps, and he takes command of it to-day. General Wagner, who has been commanding it, resumes command of his brigade, — Second Brigade of the same division."[3] This was the brigade of which Colonel Lane had recently been in command. The chagrin caused by this order of General Thomas was of course hard for General Wagner to bear, and one is not surprised to find it followed by his request to be allowed to retire from the Army of the Cumberland. On the 9th of December, in one of Thomas's Special Field Orders, a paragraph reads as follows: "Brigadier General George

[1] *Ante*, p. 79, note.   [2] O. R., xlv. part i. p. 270.   [3] *Id.*, p. 152.

D. Wagner is at his own request relieved from further duty with the Army of the Cumberland, and will proceed to Indianapolis, Indiana, and report by letter to the Adjutant General of the Army for orders."[1]

As soon as there was time for investigation, the evidence became conclusive that General Wagner's version of his orders to his subordinates and the character of the retreat from the outer line could not be sustained, and that the truth as to the facts was that which I have narrated in a preceding chapter.[2] When my official report was written, a month later, it had necessarily to be in accord with the facts as we had then learned them. The inconsistency between this and the preliminary report in regard to the points on which I had at first accepted General Wagner's statement is not a matter of regret, because it makes record evidence of my desire to take the most friendly and favorable view of a comrade's conduct while it was possible to do so.

It is right that we should even now recollect that Wagner recovered his balance before the two brigades had come to blows with the enemy, and that he then gave orders consistent with those he had received.[3] The period of over-excitement, however, had lasted long enough to prevent the correction of the error. His report was not wholly erroneous as to his commands. It omitted, as he would have been glad to have his subordinates omit, the contradictory orders which in his excitement he had issued. General Thomas was forced to conclude that this loss of self-possession showed that he was overweighted with the command of the division, especially as the brigade commanders had evidently lost confidence in his capacity.

[1] O. R., xlv. part ii. p. 117.   [2] *Ante*, chap. vii.   [3] *Ante*, p. 107.

General L. P. Bradley had been the permanent commander of the brigade which was under Colonel Conrad at Franklin, the latter having taken command by seniority after Bradley was severely wounded at Spring Hill, the day before. At the close of the war General Bradley was commissioned in the regular army in recognition of his services in the volunteers. In a letter to me, dated November 13, 1889, he says: "I was not in the fight at Franklin, as you will remember, but was in the town when the battle was being fought, and soon after saw and talked with officers who were wounded there. There has never been any doubt in my mind since then as to the responsibility for the exposed position of the two brigades of the Fourth Corps in front of the lines. It was one of the vagaries of Wagner's mind that an assault in force should be resisted by the pickets in front of a fortified line, and I remember a difference I had with him at Columbia, where it was thought we might be attacked when I was in charge of the picket lines. I felt justified then in saying to him that if Hood's army attacked, I should retire the pickets after giving information of the enemy's movement. Wagner distinctly ordered the brigades to remain outside the line and fight, after he was informed of the approach of Hood's columns by one of Conrad's staff officers. His orders and conduct at Franklin justified his removal from command of the division."

It will at once be seen that this also throws strong light on the difference between Colonel Opdycke and General Wagner in regard to this very point of risking a serious engagement in front of our main lines.[1] Opdycke practically took the responsibility of dis-

[1] Ante, pp. 73, 226.

obedience by marching his brigade to the position
reserve, which proved so important to us. He
in fact what Bradley had notified Wagner he wo
do in a similar juncture. That Colonel Conr
Bradley's successor, shared this judgment of
two senior brigade commanders, his report leaves
room for doubt. Lane's report, though made af
Wagner's removal from the command of the div
ion, is in accord with Conrad's as to the princi
facts. It is thus made plain where were the dissat
action and want of confidence which General Thon
judged to be well founded.

CHAPTER XVIII

DOUBLE BREASTWORKS ON CARTER HILL

The Two Lines at our Right Centre — Conflicting Memory of Eyewitnesses — Confederate Testimony — Solution of the Matter — Bullet Marks on Brick Smoke-house — Summary of the Evidence.

AMONG the questions at one time warmly discussed was the construction of the second line of earthworks and barricade near the Carter house, continuous with the retrenchment across the Columbia Turnpike, and the office and smoke-house of the Carter place.[1] This was the line held by Opdycke's and Strickland's brigades, after the break at the centre was restored, as we have seen; and it was through this second line that Colonel Bond passed with the 112th Illinois in the effort to regain the original front line at the locust grove.[2]

In my official report, I spoke of this matter in these terms: "The condition of the atmosphere was such that the smoke settled upon the field without drifting off, and after the first half hour's fighting it became almost impossible to discern any object along the line at a few yards' distance. This state of things appeared to have deceived Colonel Strickland in regard to his line, as he reported the first line completely reoccupied along his entire front after the repulse of the enemy's first assault, whilst in fact a portion

---

[1] *Ante*, pp. 56, 117.   [2] *Ante*, p. 161.

of it at his left was not filled by our troops, and Colonel Opdycke, not being personally acquainted with the lines, was not aware for some time that he had not reached the first line in Colonel Strickland's front, where the outbuildings of Carter's house prevented the line from being distinctly seen from the turnpike, even if the smoke had not formed so dark a covering." [1]

The only inaccuracy in this report was that in the smoky air and the quickly advancing twilight, I underestimated the length of this second line of Strickland's, and thought his confidence that he held his original front was justified as to his right, where the 72d Illinois was placed. The curve of the lines toward the rear made them appear to unite.[2] But we have seen that the right wing of that regiment, which had not been involved in the rush of Lane's brigade from the front, was ordered back to the second line, and the left flank of Moore's brigade was covered by the companies of the 101st Ohio, which were sent by General Kimball in obedience to my order.[3] At this point, however, the distance between the lines was very small.

When the opportunity came to examine the field again in December, on our advance after the victory at Nashville, the exact extent of the break from the first line was still uncertain, for the loose barricade of rails and logs in front of Strickland's right had been scattered, or used as fuel by encamping troops that had twice traversed the ground. The solider parapet, which had been occupied by the 44th Missouri, was there, and its length was that which I took to be the extent of the whole second line, or nearly so. All this has since been satisfactorily settled by

[1] O. R., xlv. part i. p. 354, and Appendix B.
[2] See sketch, p. 43, *ante.*   [3] *Ante,* pp. 131, 134.

the comparison of Confederate reports and accounts with our own, and by the statements of Mr. Carter, the proprietor of the farm, who subsequently levelled the earthworks themselves.

Yet some of the most able and intelligent of the officers and men who fought there continued to believe that the charge of the reserves at the centre had carried our line up to the original front line of earthworks on the right of the turnpike, as it certainly had done on the left. Colonel Opdycke strenuously insisted upon this as long as he lived, and in a very interesting correspondence with me on the subject in 1880, maintained the same view. Major A. G. Hatry of the 183d Ohio, which was part of Strickland's reserve, is one of the officers to whom I am indebted for valuable reminiscences of the campaign; yet he also, in reply to inquiries on this point, had no recollection of the second line, and believed that the line first constructed along Strickland's brigade front was that on which the brigade rallied and fought until it was withdrawn at midnight. More reputable witnesses it would be impossible to find; and yet the fact that the Confederates continued to hold the outside of the parapet near the locust grove is so thoroughly established that one feels impelled to seek a solution of this problem of varying memory among men of perfect honesty of character with full opportunity, apparently, of knowing the facts.

In the case of Colonel Opdycke, it seems to me that the explanation is found in his position as commandant of a reserve which lay on the slope of the knoll some two hundred yards or more in rear of the Carter house, the houses and trees as well as the roll of the ground hiding the breastworks at which our men stood. As his brigade deployed and advanced driving the enemy before them, he found the Confederates holding the outside of the retrenchment which we had built across the turnpike in rear of the opening where the main line crossed the road. This retrenchment, it must not be forgotten, was abreast of and in line with the wooden office building and brick smoke-house in the yard of the Carter house.[1] West of these was the épaulement for the 20th Ohio Light Artillery, and still farther to the right the breastwork of the 44th Missouri Regiment in the second line. This line of buildings and works was ablaze with the enemy's musketry fire, and was the objective for Opdycke's men as they charged forward on both sides of the Carter house.

From the turnpike, where Opdycke was in person, nothing could be seen of our works beyond, even if the air had been clear; but in the smoke of battle, the line of retrenchment connecting with the buildings and the Ohio battery seemed so naturally the original main line that he assumed it to be such with undoubting confidence. His mention of the recaptured battery shows that he thought it was in the original front line.[2] The successive lines of the enemy coming forward in headlong assault allowed no time for investigation, even if there had been anything to suggest it, and the desperate fight to hold the works thus gained continued on late into the night. West of the turnpike the ground fell away more rapidly, and the main line was on a level so much lower than the house and outbuildings, that it could not be seen at all from the turnpike in rear of the retrenchment crossing the road, where the brigade headquarters were during the fight.

[1] See sketch, *ante*, p. 43.
[2] O. R., xlv. part i. p. 240.

Besides the overwhelming personal evidence which fixes this second line, there are to be seen to this day the bullet marks on the east side of the brick smokehouse. The wooden office building was nearest the turnpike, and the men who stood in the space between the two buildings fired obliquely to the right, in the evening, to reach the Confederates who still held the outside of the main parapet at the locust grove. In the darkness they sometimes obliqued too much, and their bullets struck the brick wall. The leaden missiles left a glaze upon the brick surface before penetrating enough to be stopped or to rebound. The wall is dotted with these peculiar comet-shaped marks, the groove deepest toward the enemy, with the film of lead adhering where the bullet first touched. There is thus a demonstration that the firing was outward from the line of these buildings.

Major Hatry's experience in the battle was such as to account easily for his being misled on this point. He had been field officer of the day in command of General Ruger's skirmishers, and they had only come in with Wagner's two brigades when these were driven back. He came through our lines at the Columbia Turnpike, and, separating himself from the crowd of Wagner's men that were surging along toward the village, he rejoined his regiment on foot, and found them in the rifle pits, fighting the Confederates hand to hand. Lieut. Colonel Clark of the regiment and a number of other officers had been killed, and he found his own work fully cut out for him. He states explicitly that the regiment remained where he found it till it was ordered out at midnight.[1] As the line

[1] From a written statement, made by Colonel Hatry in January, 1894, in which he kindly answered specific inquiries made by me. He is a retired business man of Pittsburgh, Pa., of high standing in that city.

had been re-formed before Major Hatry joined the regiment in the turmoil of that furious combat, we can easily comprehend the impression on his memory, when at a later time the subject of the two lines was broached, that this was "news to him."

There is scarcely an engagement recorded in which there are not similar discrepancies of recollection; for in the midst of fierce excitement the thing one is specially occupied about is so absorbing that it frequently occurs that no mental note is made of other things happening under one's eyes. Remoter events, outside the range of immediate duty, are as if they had never been. It is the concurrence of positive testimony from different and independent sources which establishes beyond dispute such a fact as the construction of the second line. The lack of knowledge or of memory can hardly compare in weight with the distinct and affirmative statement of an act by one who took part in it. Such a statement as that of Captain Bates[1] of the 125th Ohio (in Opdycke's brigade), that his men constructed "new barricades," greatly overweighs, even when standing alone, the absence of recollection by another. But when we find the same fact reiterated by numerous eyewitnesses whose reports are unknown to each other and are sent up through different channels, — when both National and Confederate reports agree in regard to the matter, — when the official map of the field made by General Schofield's chief engineer shows the artillery epaulement in the second line which was extended to the right by the breastworks of the 44th Missouri, the question of fact must be regarded as settled, and the only problem remaining is the interesting but much less important one touch-

[1] *Ante*, p. 116.

ing the lack of memory in the case of those whom we should expect to know.

A brief summary of the affirmative and independent statements from Union officers who knew of the existence of the two lines will show how surprisingly strong is the support for the narrative of my official report and of my historical volume, for which, of course, my personal knowledge was the original basis.

General Ruger's official report states that "at the first onset of the enemy, the left of the line, held by the 50th Ohio Volunteers and the 72d Illinois Volunteers, fell back some fifty yards from the breastworks, at which position they were rallied and maintained a firm stand, holding this new position, which was hastily intrenched during the intervals of the fighting."[1]

Beginning at the right of Strickland's brigade, the official report of the 72d Illinois, made by Captain Sexton, describes the disorderly retreat of the outpost brigades with the enemy upon their heels. "At the same time," it adds, "the support on our left gave way, and the flank of our regiment being turned, the four left companies fell back, and as our right flank also became exposed to the enemy the remaining companies were also ordered to retire to the second line of works, which was done."[2] All the field officers being wounded, Captain Sexton took command. He ordered an effort to regain the first line, which was unsuccessful, the whole color guard being shot down in the charge across a space of some twenty yards.

The 44th Missouri built its breastworks in the second line and held them throughout the battle,[3] being the nucleus on which the first line rallied. Its official report to the Adjutant General of Missouri also de-

[1] O. R., xlv. part i. p. 365.   [2] *Id.*, p. 393.   [3] *Ante*, p. 117.

scribes the rout of Wagner's men, saying they " retreated in great disorder and confusion, literally running over the 44th, who, notwithstanding the shock, stood firm. . . . About sunset we received orders to charge over the works and retake the lower ditch, out of which the enemy had driven several of our regiments in the first charge. . . . Here Colonel Bradshaw fell, pierced with seven balls, but fortunately not killed. . . . We were forced back to our old position without being able to carry off our dead and wounded." [1] Lieut. Colonel Barr then assumed the command and made the report.

My inspector, Major Dow, accompanied the 112th Illinois, which I sent to assist in regaining the first line after dark. His description has already been quoted, and is explicit as to the relation of the two lines.[2]

Lieut. Colonel Bond, who led the 112th Illinois, is equally clear in his statement of the position of the two lines.[3]

General J. S. Casement, in a letter of December 26th, 1881, giving recollections for which I had inquired, and replying to my question whether he knew anything of the second line or barricade on the right of the Columbia road, said: " I was at that barricade with you that night, and recollect how anxious you were to have the first line regained so that the rebels might not be so near to us when we withdrew." In that connection he mentions the combination of a sally from the left of the turnpike with the direct advance from the second line.

Captain L. T. Scofield,[4] topographer, in a letter of December 23d, 1870, giving his personal memories of

[1] Report of Adjutant General of Missouri, 1865, p. 276.
[2] Appendix E.   [3] *Ante*, p. 162.   [4] See *ante*, pp. 103, *et seq*.

the field, said: "The first brigade on the right of the pike did not stand so well [as the line on the left], for they were partly made up of new troops, and the fortifications were not as strong, which resulted in their being compelled to fall back to the second line, where they were reinforced by Opdycke's brigade of Wagner's division."

The 36th Illinois Regiment (of Opdycke's brigade) published a regimental history, in which it is said that "the charge of the first brigade [in which they were] was not entirely successful in regaining the whole line at the first onset," but that "a small salient to the right of the Columbia pike for a short time was held by the enemy, who determined to use it as an entering wedge through which to break the Federal line and recover the works."[1]

The report of Captain Bates for the 125th Ohio, also of Opdycke's brigade, that they constructed new barricades, with the 36th Illinois on their right and the 24th Wisconsin on the left, need not be repeated in full.[2]

The 24th Wisconsin, of the same brigade, in reporting to the Adjutant General of that State, says that the charge succeeded in "retaking a part of the line which had been momentarily held by the enemy."[3]

Colonel Sherwood, whose regiment (the 111th Ohio) was at the left of Moore's brigade, and Captain Dowling, the brigade inspector, are explicit in stating that Strickland's men occupied the second line during the hottest of the fight.[4]

---

[1] History of the 36th Regiment Illinois Volunteers, by Bennett and Haigh, Aurora, Ill., 1876.
[2] O. R., xlv. part i. p. 251.
[3] Report Adjutant General of Wisconsin, 1865, p. 372.
[4] *Ante*, p. 131.

Mr. Carter, the present proprietor of the Carter place, levelled the " new barricades," and pointed out to Captain Speed and others the site of this second line as he found it when the battle ended.[1]

Captain Twining's official map shows the artillery epaulement in rear of the Carter house, which was in the 44th Missouri line, as a second line in rear of the main intrenchment on the right of the turnpike. Major Foster's Confederate map, made for Stewart's corps, also shows the double line at that point.[2]

Thus from more than a dozen independent sources comes the multiplied confirmation of the statements of my official report, without mentioning the reports of the Confederates, which are to the same effect.

[1] Ohio Loyal Legion papers, vol. iii. p. 61.
[2] *Ante*, pp. 45, 83.

CHAPTER XIX

THE RALLYING OF THE OUTPOST BRIGADES

Value of Detailed Reports — Landmarks — Lines of Retreat from the Outpost — Crowding toward the Centre — Significant Omissions — Lists of Missing — What they Teach — Reports Compared — Incidents relating to Captured Flags — Statements of various Officers — Conclusions from the Facts.

ANOTHER subject of debate has been the question, What became of Wagner's two brigades which were driven in from the front? I have already given the story of their break from their outer position, the rush of the main body along the turnpike, and into the town of Franklin, the effort at reorganization near the river, and their final collection near the wagon bridge and withdrawal in the night under the orders to march to Nashville. I have also noted the fact that numbers of men from these two brigades turned and fought at our main lines with the brigades of Reilly, Opdycke, and Strickland which were the organized troops holding the centre on both sides of the Columbia Turnpike.

The reports of the commandants of Wagner's two brigades in question ignore the fact that the division and brigade commanders made their official rallying place in the town, and confine their statements to those disorganized portions of their commands which halted and fought at the main line. These are spoken of as if they were "the command." Nothing is more notorious in regard to military reports than that they are apologetic in cases of mishap, and no form of glossing the facts is more common than the omission of unpleasant features whilst more creditable ones are amplified.

The task of the historian will be to reach the truth by running down and fixing the facts which are omitted, and constructing an authentic narrative based upon everything which is satisfactorily and affirmatively established. The greatest help in this work is found in detailed subordinate reports, and the help is increased just in proportion as these are full in incidents, with time, place, and circumstance which may be compared with evidence derived from other sources. With general and rather vague statements that a command rallied at the line and fought bravely for hours, we can do little except to show opposing statements that are inconsistent with these; but when details are given, as of the other troops on right and left with whom they were in contact, commands received and given, by whom and to whom, the general picture begins to assume shape that can be tested, and apparently insignificant incidents often become of great significance.

It would have overburdened my narrative to have analyzed the evidence to which I am now referring, and I have preferred to give a connected story from my own standpoint, as I knew it and had to act upon it at the time. I trust that now, however, it will not be unprofitable to consider some of the statements which may at first appear conflicting, and to weigh more fully the evidence as to the rallying of the outpost brigades.

First, as to the place in the main line where Lane's and Conrad's brigades must have crossed our works.

In their outpost position they had been in a wedge-shaped formation on both sides of the Columbia Turnpike, the road being in the apex of the wedge. Conrad was on the left and Lane on the right.[1] In our main works nearly half a mile in rear, Reilly's brigade was behind Conrad, and Strickland's was behind Lane. Opdycke's brigade was some two hundred yards still farther in rear, lying in column of regiments in reserve on the west side (right) of the turnpike behind Strickland. As to all this there is no controversy, and the places are distinctly marked in the official map of General Schofield's chief engineer, Captain Twining. The same map marks the place of the locust grove near the right of Strickland's brigade, which was partly cut down to form an abattis in that part of the line, and which is so often mentioned in the official reports that it becomes a fixed and important landmark.

These positions indicate so plainly what must occur when Wagner's two brigades should be overwhelmed and driven in by the enemy, that every narrative must be interpreted with reference to them. Conrad's and Lane's men would naturally converge toward the turnpike which was the unobstructed way to safety, the abattis in front of the works being a most inconvenient thing to cross. All the details in the reports and in authentic personal narratives that have been published show that such was in fact the line of retreat of Wagner's men. The mass crowded the turnpike on the way into the town, whilst thinner portions of the line threaded the abattis and climbed the breastworks on either side of the turnpike.[2] Lane's report mentions the "heavy line of abattis of locust boughs, placed there for some purpose, through which my line

---

[1] See map, p. 45, *ante*.   [2] *Ante*, pp. 104, 109, 118.

had to pass."[1] Not only do all the official reports show that Moore's brigade, on the right of Strickland, firmly held its place, but the statement of Mr. Gist (to which reference has already been made), who belonged to the right regiment of Lane's brigade, is explicit, that "Moore's brigade repulsed the attacks made on their part of the line," and "had not been disorganized in the least."[2] He also says that his part of Lane's line "reached the works just at the right of the Columbia pike, near a grove of small locust trees."

On the left of the turnpike there is a similar concurrence of testimony that Casement's brigade remained unmoved, and that those of Conrad's men who did not come in upon the road "were mixed from the cotton gin on to the pike."[3] It is thus put beyond dispute that whatever these two brigades of Wagner's did or suffered in the main line was between the cotton gin on the left and the locust grove on the right, comprising not quite the whole of the original front of Reilly's and Strickland's brigades, in the centre of which was the Carter house where Opdycke's brigade (also of Wagner's division) was, after its rush forward from its place in reserve, and where my personal headquarters had been since daybreak in the morning and remained till we withdrew at midnight.

If Conrad's and Lane's brigades were in the main line, as nobody doubts that Opdycke's was, then the whole of Wagner's division, except stragglers, was there. Wagner's place should have been close to my own, Conrad should have been close to Reilly, Lane

[1] O. R., xlv. part i. p. 256.
[2] Professor Gist's Narrative, referred to, *ante*, p. 201.
[3] Report of Colonel Buckner, 99th Illinois (of Conrad's brigade), O. R., xlv. part i. p. 280.

close to Strickland and Opdycke. Each of them should have been cognizant of what occurred there, have been necessarily in communication with General Stanley during the short time he was with Opdycke's brigade before he was wounded, and with myself afterward. My efforts completely to restore the line in Strickland's front, in which I called for assistance from Kimball's division on the extreme right and brought a regiment from Stiles's brigade on the extreme left, must have involved co-operation with Wagner and his brigade commanders. Of Opdycke's presence and the place of his regiments, myself and his neighbors on right and left were well aware, as both official and unofficial reports show; but none of us had any knowledge of the others or communication with them, nor do their reports show any relations to us. It cannot be necessary to point out the significance of this as bearing upon the question of their actual whereabouts, and the place of the formal reorganization of the two brigades.

Before passing to the examination of the reports of General Wagner and his subordinates, it may be instructive to notice that the reports of the "missing," in the case of such a mêlée as occurred at our centre, fix the extent to which the enemy were temporarily within our works, in a manner which strikingly corroborates the other evidence. On reaching Casement's brigade on the left and Moore's on the right, the list of missing becomes merely nominal, amounting only to a few who would straggle in the darkness. In Reilly's, Opdycke's, and Strickland's brigades, however, they are numerous enough to indicate a considerable number of wounded and other prisoners who fell into the enemy's hands within our breastworks, and were carried back with them into captivity when they were driven out of our lines. The statistical tables thus mark the extent of the hand to hand fight at the works as accurately as an eyewitness could do.

Second, I shall bring together some of the evidence from the reports and documents corroborating the account which I have given of the reorganization of Wagner's two brigades at the river.[1]

Wagner's own report, written on December 2d, and in advance of the receipt of his subordinates' reports (except Conrad's), is very vague and partly unintelligible in its reference to the rallying of his command. The language as found in the Official Records is this: "On reaching the main line of works the officers rallied their men as best they could, and placed them in position to support the works which were give up [*sic*] their position at the approach of the enemy who followed close on the steps of our retiring lines."[2] The latter part of this sentence (which is the only attempt in the report to describe the rallying) is without meaning, and shows that the writer had found difficulty in casting his statement into any form which should harmonize his own knowledge of what occurred with the report of Colonel Conrad, which alone was then before him.[3] It is significant, however, in this, that the troops are said to have been "placed in position to support the works." This implies that they were as a body at some distance in

---

[1] *Ante*, chap. x.   [2] O. R., xlv. part i. p. 232.

[3] The original manuscript of Wagner's report in the War Records Office is literally as given above. It was apparently copied from some previous draught in preparation for his signature, as there are no erasures or interlineations in the passage quoted. Its lack of connection suggests that possibly something had been erased in the draught with a purpose to change the form of expression, and that haste had caused this to be forgotten.

rear, for thus only would they be in "support" of the fighting line. The report makes no mention of the troops of other commands, or his contact with them; and the omission is in harmony with the facts as to the general rallying place of the two brigades, which have been already narrated. He had no personal knowledge of the situation on the line, and could not speak of what occurred after he had passed to the rear.

Colonel Opdycke's official report necessarily implies the total severance of communication between the brigade and his division commander from the time it was placed in position as a reserve until the withdrawal at midnight. As to this period, it makes no allusion whatever to General Wagner (to whom the report was addressed), but states that the order to be ready to charge up to the line and the directions to withdraw at midnight came from myself. In his narrative of the battle, published after the close of the war,[1] he explicitly states what is thus impliedly contained in his report, saying that "General Wagner was carried to the rear in the rush of disordered troops, and did not again find his way to the front." My own opinion that he was properly engaged in the reorganization of the broken brigades, and that this was his first duty, has already been given.[2]

Colonel Allen Buckner, of the 79th Illinois, seems to have been the officer next in seniority to Colonel Conrad in that brigade. He thus describes the rally at our works in his report: "We fell back to the works to the left of the pike, and I was enabled to rally, and afterward fought in connection with troops of the Twenty-third Corps, and others of our brigade and corps (for here we were mixed from the cotton

[1] New York Times, September 10, 1882. [2] *Ante*, p. 145.

gin on to the pike) until some time in the night. The troops being thick and we not needed longer, I told General Reilly that I would get my regiment back, and try to get off our wounded. In a short time orders came, and we got the brigade together, and came out of the town about midnight." [1]

This shows very clearly that Reilly's brigade of the Twenty-third Corps was in position at the cotton gin, with General Reilly personally in command; that Colonel Buckner and the men with him fought in that line after rallying; that Reilly's men were enough to hold the line after the first repulse of the enemy, and that because the troops were so "thick" Colonel Buckner very rightly gathered his men for reorganization; that his brigade and division commanders of the Fourth Corps were not present, so that it was to General Reilly of the Twenty-third Corps that he communicated his wish to get his men back from the line; that it was after this that "orders came" (from his own superiors), and they "got the brigade together." Stronger corroboration of my general narrative could hardly be given.

No regimental reports of the battle were made by the regiments of Lane's brigade, except the 97th Ohio (Colonel Lane's own), in command of which was Lieut. Colonel Barnes. His vivid description of the retreat of his regiment through the abattis in front of the locust grove on the right of the Columbia Turnpike has already been quoted, as well as his statement of the utter confusion and disorganization of the brigade. In concluding it he says that about ten o'clock the firing gradually ceased, and he "received an order from Colonel Lane in person to draw off the regiment and reorganize the line." "In gathering them to-

[1] O. R., xlv. part i. p. 280.

gether," he adds, "they came from the front."[1] This implies that the order was given at some place in the rear, and to it, as the place of reorganization, the group of Lane's men, who had been stubbornly fighting with Strickland's brigade, "came from the front" as word reached them that a rallying place had been established.

In response to a call for flags and trophies captured, several regiments of Wagner's division, two or three months later, made claim to the capture of flags, which, as was said, had been turned over to officers of the Twenty-third Corps. The alleged circumstances throw light on the situation.

Such a claim was made for Sergeant Ransbottom of the 97th Ohio, Colonel Lane's regiment, a remnant of which we have just seen was fighting at the locust grove in Strickland's brigade line. Lieut. Colonel Barnes says that after nightfall "volunteers were called for to pass through a gap in our works on the Columbia pike that they might enfilade the enemy, and capture a portion of the storming party."[2] Sergeant Ransbottom is said to have volunteered among others, and to have captured a flag. The only order of the kind given that evening was my own order to General Reilly already mentioned,[3] to be executed by the troops of his command on the east of the turnpike. There was no attempt to pass out at the gap on the Columbia Turnpike. That was in Opdycke's brigade line, and was quite too near the enemy, who were holding the outside of the original line in Strickland's front. The party made the movement from the salient at the cotton gin. If the sergeant volunteered for that sally, he was in the left wing of Reilly's brigade, with two full brigades inter-

[1] O. R., xlv. part i. p. 265.   [2] *Id.*, p. 267.   [3] *Ante*, p. 146.

vening between it and the locust grove, where Colonels Lane and Barnes locate the group of the regiment to which he belonged. He acted, if at all, as an individual temporarily attached to another command, and far separated from his own comrades. No call was made upon any troops on the right of the turnpike to participate in the sally mentioned. Yet the incident is referred to in the brigade report as an item in the creditable conduct of that command.[1]

In a similar report as to flags, under date of January 5, 1865, Major Atwater of the 42d Illinois (Conrad's brigade) states that different men of his regiment captured flags (three or four) or went over the works in the evening and picked them up where the enemy had let them fall, but that on their coming in they were ordered by officers of the Twenty-third Corps to give them up, and they obeyed.[2] These officers are said in his communication to have been of the 104th Ohio, which was in Reilly's brigade, and stationed at the cotton gin. If the incident is correctly reported, it shows that no officers of Conrad's command were present, for had they been, their men would have appealed to them and they would have asserted the claims of that command. The controversy would then have been between commissioned officers, and would not have rested on the statement of private soldiers alone. It shows also that the officers and men of Reilly's brigade were in orderly line attending to their duties, and that the scattered individuals of Conrad's command were acting with Reilly's regiments and under the orders of his officers. Thus from unexpected sources we get aid in establishing the true condition of things upon the line.

The narrative of Mr. Gist, from which I have already

[1] O. R., xlv. part i. p. 256.   [2] *Id.*, p. 276.

The Rallying of the Outpost Brigades 253

quoted,[1] says: "Long after night there was a lull in the battle, and we began to hope that there would be no other charge. I moved around to the right a little, on the works, and met a member of my company, the only one of the regiment that I had seen since the opening of the fight." As his regiment was one of Lane's brigade, and his own position was at the locust grove where was the only considerable group of the men of that brigade, the fact is more significant of the thin scattering of his comrades than any general estimate of their numbers could be. He tells another incident also which corroborates my statement that this part of the line was held by the 72d Illinois of Strickland's brigade.[2] "A colonel of some regiment tried hard in the thickest of the fight to get the line to charge.... He mounted the works himself and called upon the disorganized portions of a dozen commands to follow. He was pierced by a ball and fell a few feet to my left." This officer was Lieut. Colonel Stockton of the 72d Illinois, who with Major James of the same regiment was wounded and fell. Captain Sexton,[3] upon whom the command devolved, has permitted me to read a letter from Lieutenant Mohrmann of the same regiment describing the scene when Colonel Stockton mounted the breastwork, which Captain Sexton also confirms.

Captain Sexton estimates higher than any other officer of the Twenty-third Corps the number of men from Conrad's and Lane's brigades who rallied on the line, placing them at "about five hundred men." No doubt a larger body of them took temporary place in his regiment than in any other on the line. He says of them, "There were a few *line*, but no field officers with them." He saw nothing of their division or

[1] *Ante*, p. 246.   [2] See *ante*, p. 118.   [3] See *ante*, p. 118, note.

brigade commanders. As to the place where the bulk of the two brigades were rallied, he says: "I was informed that part of them were stopped by Wood's command near the river, and the rest at Nashville."[1] His narrative is full of admiration for the soldierly conduct and courage of those who rallied in his line, and for the cool discipline they showed in maintaining the fire along with his own men.

Major Hatry, of the 183d Ohio (also of Strickland's brigade), we have seen[2] in command of the skirmish line of Ruger's division which came in with Wagner's broken brigades. These, he says, in his written reply to questions from me, "crowded the Columbia pike at the Carter house." Leaving them he went to his own regiment which he found in line with the 44th Missouri and 72d Illinois, the centre and right of Strickland's brigade, holding breastworks which, as he says, they did not move out of " until we evacuated them at midnight." As to the men of Conrad's and Lane's brigades, his statement is: " I don't think many stayed with us in the front line. They proceeded down the village, and were rallied down there by some officer whom I saw come up with reinforcements. How many came back, and where they went, I could not say, but think that some of them went to both sides of the pike or very close to it in the rifle-pits there." To the distinct question whether he saw Wagner, Lane, Conrad, or any of their regimental commanders there, he replies, "I did not." He clearly identifies his position on the field. "The line we fought in was in front of the locust grove and to the right and front of the Carter house, I mean as the ground descends from the

---

[1] From Captain Sexton's written statement referred to at p. 163, *ante.*

[2] *Ante,* p. 237.

house. Our regiment and the part of the line in which they were was near the rear of the Carter house and to the right of the locust grove." [1]

The graphic description by Captain Scofield of the retreat of the two brigades has already been given,[2] but in the same correspondence he adds what is directly in point touching the organization of the line when it was restored. "I did not see," he says, "any superior officers commanding a brigade, except Reilly, Opdycke, and Strickland, between Casement's right and Moore's left, and I am sure there were none there, for I should have seen them, as I was not out of sight of Columbia pike from the time the battle commenced till 10 o'clock P. M." Again, he says: "There were no detachments from the retreating brigades, in command of officers, brought to our line during the action. There may have been stragglers from the rear that returned to our position singly or in squads that would escape my notice, but anything like a regiment or company I could not have missed seeing."

Major Dow, my division inspector, in his written statement says, "I don't remember seeing Wagner during the battle."[3] The visit of my Adjutant General to General Wagner at the place near the river where he was reorganizing the two brigades need not be repeated.[4] In a correspondence with General Stanley to which I shall have occasion by and by to refer,[5] he recognized the fact that Opdycke's brigade remained under my command till the midnight withdrawal, separate communication being had by him and Wagner with the other two brigades of that division

---

[1] See map at p. 45, *ante*. The slope from the Carter house is here toward the southwest, and the line followed the curve.
[2] *Ante*, p. 104.
[3] See Appendix E.
[4] See Appendix F.
[5] *Post*, chap. xx.

of his corps. To any one having the slightest knowledge of military organization this is completely destructive of any theory that Conrad's brigade was on one side of Opdycke and Lane on the other, the whole under the usual division organization. In such a case the withdrawal of Conrad and Lane in the night leaving Opdycke without orders or communication with his superiors would be a military absurdity. It could only have been done by full communication between such superior officers and myself, with careful precautions to prevent mischievous results on the line. Nobody has ever pretended that this was the case. The fact that Opdycke alone remained under my orders and was withdrawn by me, as he officially reported, would of itself prove that the other brigades were not there.

Thus from every source, affirmative and negative, the mass of evidence establishes the historical facts as I narrated them from my personal knowledge. Various motives of pride and anger led the superior officers of that division to omit the report of the two brigades reassembling and reorganizing at the Harpeth River, and to treat the groups which rallied with the organized brigades upon the line as if they were the whole command. I think the candid reader will see how futile such a treatment of the subject is, when once the mass of evidence official and unofficial, direct and circumstantial, is considered. But I again repeat that after the two brigades had been compromised and broken in front, it was no fault of Wagner or his subordinates to rally for reorganization wherever the crowd surging along the Columbia Turnpike into the town could be stopped. To get them into form so that they could be handled was their first duty as officers. Those who stopped at the line obeyed a noble impulse

The Rallying of the Outpost Brigades 257

in doing so, but it was their manifest duty also to find the *cadres* of their regiments and brigades at the earliest moment when it was evident that the organized brigades of Reilly, Opdycke, and Strickland were holding the position.

With the facts before us there is little use in trying to estimate the numbers of Lane's and Conrad's disorganized men who halted at our breastworks and fought. General Reilly was prompt to give credit to the 175th Ohio[1] (a new and unassigned regiment), and to the 44th Illinois (the left wing of Opdycke's brigade), which gave him organized help in regaining and holding his breastworks. There is not the slightest ground to doubt the equal candor of his report that "with this exception the brigade received no assistance during the fight, unless perhaps some of the men coming in over the works may have rallied in or behind the lines." Colonel Opdycke assured me that only detached individuals or small groups rallied with his men. In Strickland's line it appears that considerably larger numbers rallied at the locust grove, but, like the rest, completely disorganized. At this point alone does my own memory recall anything like a collection of Wagner's men going to join their brigades in town. Seeing in the evening what I took to be about a company dimly outlined in the darkness passing, at a little distance, my own position on the turnpike near the Carter house, I directed a staff officer to inquire where they were going, and was informed that they were some of Wagner's men who had been collected and were going to join the command in the village. Others had doubtless done the same, but in groups too small to attract attention.

[1] O. R., xlv. part i. p. 412.

CHAPTER XX

AN UNEXPECTED CONTROVERSY

Colonel Stone's Paper in Century War Book — General Stanley's Criticism — A Violent Attack — Earlier Correspondence — Nine Points — Two Corrections — Basis of a Historical Narrative.

THE matters of more or less debate which I have discussed were such as involved no special interest on my part, beyond that which is felt by every actor in important events, and by every student of history who desires to solve correctly the problems of a great war. I must now, however, give some attention to a controversy which more nearly concerns myself. I would gladly be excused from this; but historical truth is involved, and the discussion of the battle of Franklin would be incomplete without it.

In the series of important papers upon the Civil War published in the Century Magazine, and afterward collected into the Century War Book,[1] was one upon the campaigns of Franklin and Nashville by Colonel Henry Stone, formerly upon the staff of General George H. Thomas. In this paper Colonel Stone had said of the main line in front of the village of Franklin, that "all the troops in the works were ordered to report to General Cox, to whom was assigned the command of the defences." To this statement of Colonel Stone, made entirely upon his

[1] Battles and Leaders of the Civil War, Century Co., 1884-88.

independent investigation, General Stanley took exception, in a letter which was published in the Century Magazine for February, 1889. He asserted that the statement of Colonel Stone was not made "from the standpoint of an officer well informed as to the rights of command," for "all the troops in the works could not have been ordered to report to General Cox without removing me [Stanley] from the command of the Fourth Corps, and no one will claim that the latter idea was ever thought of by any one." Certainly no one has been more explicit than myself in saying that General Stanley was in command of the Fourth Corps, and continued so.

This, as a criticism of Colonel Stone's paper, whether well or ill made, would have involved me in no way, had General Stanley not added an ostensible excuse for Colonel Stone, "as he could easily have been led into making this misstatement by General Cox himself; for the latter, in the book written by him, entitled 'The March to the Sea, Franklin and Nashville,' on page 86 complacently styles himself 'Commandant upon the line.'" General Stanley then proceeded to controvert this assumption.

By the courtesy of the editors of the Century, an opportunity was given to Colonel Stone and myself to reply in the same number of the magazine, and I was quite content to leave the matter where it was thus placed. In the following autumn, however, a new attack appeared in a daily newspaper, which did not come to my notice till a fortnight after its publication. Foiled in an effort to put a reply before the readers of the same journal, I allowed the subject to await that clearing up of controversies which I have had an abiding confidence would follow the completed publication of the Official War Records. I was the more easily led to do thus because the matter, tone, and style of the last publication mentioned were not such as one ordinarily feels the need of replying to.

In an attempt at a somewhat full historical investigation, however, it would manifestly leave the discussion incomplete if the question of command on the line did not receive the same attention as other questions. I shall try to treat it as a simple historical inquiry, in which the evidence, both official and other, seems to me pretty complete and clear. To let it fall below this level would be to forfeit the respect of intelligent readers.

When I was preparing, in the summer of 1881, to write the volume in the Scribner War Series of Histories, entitled "The March to the Sea, Franklin and Nashville," General Stanley opened a correspondence with me, offering to assist me by the loan of papers, etc. In my reply, dated August 24th, accepting the offer, I took the opportunity to compare our recollections of the principal facts. I wrote: —

"Let me state a few consecutive points within my own memory, and ask you to compare it with yours, premising that I have not yet begun the systematic review of the documents in my possession.

"1. Two divisions of the Twenty-third Corps were present, and acting under my command, Ruger's on the right of the Columbia pike, and my own (Reilly commanding) on the left.

"2. Schofield had only intended to cover the crossing of trains, and had not meant to fight south of the Harpeth. He had therefore ordered me to send my own artillery and wagons over the river early, and had arranged that Major Goodspeed,[1] your chief of

[1] This should be Captain Bridges: Major Goodspeed was Chief of Artillery of the Fourth Corps a little later.

artillery, should detail some batteries as your troops came in, and they reported to me.

"3. After putting my own command in position, I reported to General Schofield that my troops were not sufficient to reach the river on the right, and that flank was consequently exposed. Kimball's division reported to me, and was assigned that place.

"4. I received a written dispatch from General Schofield saying that two brigades of Wagner's were out as rear guard, and one (Opdycke's) would report within the lines to act as my reserve; that Wagner was ordered to bring the other two brigades in whenever Hood showed a purpose of serious attack. I showed this note to Wagner, and found he had such orders.

"5. When Hood formed and advanced, Wagner did not order in the two brigades, but ordered them to fight. One of my staff, still living, heard him send the order from the Carter house. In his excitement, he had forgotten his orders, apparently, and did not change though reminded of them.

"6. Being at the left of the line on the parapet, watching the enemy's advance, I was amazed to see Wagner's two brigades open fire. They were quickly run over by the enemy, and came back in confusion.

"7. I immediately sent an aid to Opdycke to warn him to be ready to advance in case of a break at the centre, and to order the commandants of brigades, etc., to withhold their fire till Wagner's men should get in. The two aids who were with me are both dead, one being killed while performing part of the above duty. Opdycke afterward told me that he got no order, and acted on his own judgment, and I have accepted that as the fact.

"8. I almost immediately followed my order, and

rode to the pike. There I met Opdycke advancing, and met you also. We all went forward together. When Opdycke reached the parapet, you and I were trying to rally the fugitives immediately in rear of the line. While thus employed you were wounded, and your horse was also hit. You asked me to look at the hurt, and I urged you to go and have surgical attention to it. I dismounted Captain Tracy, one of my aids, and gave you my horse which he was riding. To say anything here of the impression your conduct made on me would violate the old maxim about 'praise to face,' etc.

"9. Opdycke and the artillery continued to act under my orders till we left the lines at midnight. Orders to the rest of Wagner's division and to Kimball went from your headquarters, you continuing in command of the Fourth Corps till we got back to Nashville, notwithstanding your hurt.

"As I have said above, I have not yet begun the collation of documents; but I have taken advantage of your kind letter to give the above outline, and to ask for any illustration, correction, or addition which may occur to you, so that I may give careful attention to any point on which my memory should differ from yours."

To this General Stanley replied from Fort Clarke, Texas, under date of October 17, 1881, saying, among other things: —

"The nine points submitted in your letter are, to the best of my memory, exactly correct. I think it may be true that Opdycke did not receive your order. When I arrived at the left of his brigade, the men were just getting to their feet, as they had been lying down, I presume to avoid the enemy's bullets."

Replying on the 31st of October to this letter,

I said: "It gives me sincere pleasure to know that we agree in memory as to the outline of the battle of Franklin. It increases my confidence in my recollection, and will give me more assurance in going forward."

This outline, so explicitly agreed upon, needs correction in but two points, in both of which the official reports show that I understated my own command upon the line. First, Opdycke's official report[1] shows that he received my order mentioned in the seventh of the points stated "to be ready to advance in case of a break at the centre," and that it was the second order mentioned in my official report,[2] calling on him to charge at once, which he anticipated. Finding my aid, Captain Tracy, with him when I joined him on the turnpike, I of course assumed that the order had been delivered, and did not know till Opdycke afterward told me of it, that the beginning of the actual charge was spontaneous on his part, and made it unnecessary for Tracy to deliver the order when he reached the brigade. The evidence that Opdycke recognized my authority was thus officially complete. In a correspondence with him at about the same time as that with General Stanley, he very frankly and clearly put the matter on this footing, as he did in his published narrative of the battle.[3] General Stanley's assent to the points recognizes also my right to exercise the command over that brigade and the artillery, and to continue it till the midnight withdrawal.

Second, in the ninth point I understated my exercise of authority as to Kimball's division, for that officer's official report[4] shows that I called upon him to detach a regiment to reinforce the left of Moore's

[1] O. R., xlv. part i. p. 240.
[2] See Appendix B.
[3] See *ante*, p. 96.
[4] O. R., xlv. part i. p. 178.

brigade, and that Colonel McDanald's was sent in response to the order.[1]

In my history of the campaign, I had followed the outline thus explicitly assented to by General Stanley. When it is remembered that this was the result of his own offer to help me fix the more important facts of the campaign for the expressed purpose of a historical narrative, their conclusiveness upon him is indisputable. If there were anything which he should afterward wish to modify or question, a courteous and temperate tone was certainly demanded as well as a recollection of the points which he had so emphatically indorsed.

So far as my use of the phrase "commandant upon the line" was concerned, I said, in my reply in the Century, that it was used chiefly to avoid the repetition of my own name in a narrative written by myself, and that I should be quite content to have the reader substitute my name for the phrase. I might have added that in my desire to keep upon ground where I supposed we were agreed, I had spoken of myself as "commandant upon the line" only in the events prior to General Stanley's appearance at the front, and my language after that was, "the commandants of the two corps met on the turnpike just as Opdycke and his men were rushing to the front."[2] Controversy

---

[1] When the correspondence printed in the Century Magazine for February, 1889, appeared, I had not seen General Kimball's official report and had forgotten the fact that I had sent the order. Colonel Stone, however, referred to it in his reply to General Stanley, and I then procured a copy of the official report. I discovered, a little later, in my files of preserved letters, one from Mr. Edward C. Russell, a hardware merchant at Corning, Iowa, dated January 4, 1876, in which he recalled himself to my memory as one of my orderlies at the battle of Franklin, and the messenger who carried my order to Kimball for the reinforcement at the centre.

[2] Franklin and Nashville, etc., pp. 86, 89.

on that point was therefore wholly gratuitous. But the challenge of my accuracy went the whole length of our relations to the line engaged in the defences of Franklin at all times during the 30th of November.

My well meant purpose to keep within the lines of uncontroverted facts in my history was thus, some seven years after its publication, disappointed. I had in fact accorded to General Stanley more than he had any claim to, and the outcome was that I found my own candor and accuracy assailed.

In writing to me after the appearance of the little volume referred to, General Schofield, in November, 1882, spoke strongly of my manifest effort, in treating the whole campaign, to avoid unnecessary criticism and to deal liberally with all who had a responsible part in it. He even suggested that I might "feel estopped" from a more rigid analysis "by the generous treatment you have already given" to others. If the word "generous" be properly used in regard to anything I have said of that campaign, I am very sure it would nowhere be more appropriate than to my references to General Stanley, from my preliminary report in the field to my last printed word. Whether well or ill requited, I should not have modified the creditable picture which I had helped to draw. He has himself made it seem necessary to examine more critically the relative parts we bore in that day's work. The fulness of the narrative in the preceding pages will enable me to make the summary comparatively brief.

CHAPTER XXI

CONTROVERTED POINTS

Work assigned the Twenty-third Corps — Detachment of Fourth Corps Batteries — Orders to Ruger and Kimball — Detachments often Necessary — Articles of War and Regulations — Questions of Command — Stanley on North Side of River — His Ride to the Front — Soon Wounded — He Retires — Statements of Officers — At the Field Hospital — At Schofield's Headquarters — Summary — Official Reports — Analysis of Stanley's — Contemporaneous Records — Conclusion.

In an early chapter [1] I have told how, before the break of day, General Schofield came to the Carter house, upon the Columbia Turnpike, and gave to me in person the direction to put the Twenty-third Corps in position, make defensive works, and hold back Hood at all hazards. With the Columbia road as a centre, the line as first laid out extended about equal distances to right and left, say half a mile each way, reaching from the Harpeth River on the left to the Carter's Creek Turnpike on the right. This line was occupied and intrenched wholly by the two divisions of the corps. This was "the line" in front of Franklin during the forenoon. Even the artillery in it would have been the batteries of the Twenty-third Corps, except that General Schofield as a wise economy of time ordered me to send my artillery at once over the Harpeth, promising to detail batteries from Stanley's Fourth Corps, and

[1] *Ante*, chap. iii.

give them to me to supply the place of mine. About this there has never been any question: every report, official and private, is in accord, and General Stanley has said that it is "exactly correct."

The work of holding back the Confederates in front of Franklin was thus definitely committed to me, and Hood was advancing from Columbia by the road which cut the centre of the line thus established. As to the artillery, General Schofield, in accordance with his promise, made a detachment of certain batteries from the Fourth Corps to fight in such positions in my line as I should direct. To say that they did not come under my general command would be a military absurdity. There neither is nor for the past century has there been an army in Christendom in which a corps commander has not control of the batteries assigned to him. A reserve is often separately massed for a special purpose, but the batteries detailed to a corps are always part of "the command," for the time at least. The Twenty-third Corps had no reserve artillery, but the batteries were integral parts of the divisions; and when their place was supplied by others, as in this case, the substitutes came, by the fixed rule of the service, under the orders of the general officer to whom they were ordered to report. Any other rule would make utter confusion.

General Stanley recognized this in its full effect, since, for the very purpose of settling points as to the battle for a published narrative, he admitted the exact correctness of my statement that the batteries were "detailed" to report to me generally, and that they "continued to act under my orders till we left the lines at midnight."[1] General Schofield's pur-

[1] *Ante*, p. 262.

pose that they should so report to me and act under my orders is too plain for discussion. He had said so to me orally, and the written order from his headquarters was meant to do this, neither more nor less. Now what was the language in which he expressed the purpose? It ordered Captain Bridges to "report four (4) batteries from your command to Brig. Gen. J. D. Cox for position on the line." It is thus conclusively proved that he and his adjutant general regarded that language as aptly expressive of the purpose.

But the officer to whom it was addressed also understood it so. Captain Bridges states in his official report that he placed the Fourth Corps batteries in the positions in the Twenty-third Corps line "by direction of Brigadier-General Cox," and describes them in detail. Then, after describing the opening of the battle and the fighting of the artillery, he adds: "Receiving orders from Brigadier General Cox, commanding the Twenty-third Army Corps, and Lieut. Colonel Schofield, Chief of Artillery Department of the Ohio, to remain upon the line with the batteries, I remained near the Columbia pike, which seemed to be the place upon the line where the enemy made the most effort to obtain a lodgment, and which was a position from which I could see the entire line."[1] From the first of these quotations to the last, including the whole action of the artillery upon the line, he makes no mention of General Stanley whatever, although, as chief of artillery of the Fourth Corps, he was attached to that officer's staff. It should be noted also that his position was close to my headquarters on the turnpike near the Carter house, where Stanley was

[1] O. R., xlv. part i. pp. 320, 321.

for a few minutes in the first mêlée, and where he would have been likely to be had he remained at the front. When Captain Bridges in the succeeding sentence speaks of the order to withdraw the batteries, he correctly says [1] that it came from General Stanley. The withdrawal orders we have already examined.[2] The report of Captain Bridges thus becomes proof not only of his recognition of my own command upon the line, but that General Stanley was not there long enough even to raise a question as to subordination or command there.[3]

General Ruger of the Twenty-third Corps was also ordered to report to me, and in this case also there is no possible dispute as to the fact or as to General Schofield's purpose to do it. It was part of the authority and instruction he gave me orally in the early morning, and, like that in regard to the batteries, was a little later put in form as an order. In this, as in the other case, General Stanley admits the exact accuracy of my statement,[4] that Ruger's division was thus put under my command for the day, and General Schofield's official report declares that I was temporarily in command of the Twenty-third Corps.[5] Again we have to ask in what language was this purpose expressed in the order, and General Ruger answers the question. In his official report of the battle he says he was ordered to report to me "for assignment to position."[6] This not only shows that the same language was used as in the case of the batteries, but, by his unquestioning recognition of my authority, General Ruger shows that he

[1] O. R., xlv. part i. pp. 320, 321.
[2] *Ante*, p. 169.
[3] *Ante*, p. 100.
[4] *Ante*, p. 260.
[5] See Appendix A.
[6] O. R., xlv. part i. p. 364.

regarded the words as appropriate to convey the meaning intended.[1]

When, later in the day, as we have seen,[2] the appearance of the enemy's cavalry on my right flank showed the need of strengthening that part of the position, I reported the fact to General Schofield. He ordered General Nathan Kimball, commanding a division of the Fourth Corps, to report to me, and I placed him on the recurved extension of my right. For the third time that day the language of the order was, "The Commanding General directs that you report with your command to Brig. Gen. J. D. Cox for position on the line to-day." General Kimball not only took the position assigned him, but in the first hour of the battle, while it was yet broad daylight,[3] he obeyed my order to send a regiment to the centre, where, between Strickland's right and Moore's left, help was sorely needed.

Of these orders, thus successively sent and thus successively obeyed, General Stanley argues that they conveyed no authority except such as an "aid-de-camp" or "an orderly or guide" might have had, "to show General Kimball where he was to go"![4] Such interpretation would discredit a freshly commissioned subaltern. He goes further, and in ignorance of the fact that General Kimball's official report contradicts him, completes the absurdity by saying that, if I "had attempted to assume the authority to give orders," his division commanders "would have paid no attention to them."[5]

The truth is that no rule is more rigidly enforced in our army than that an officer is never ordered to

---

[1] General Thomas H. Ruger is now (1896) a Major General U. S. A.
[2] *Ante*, p. 60.
[3] *Ante*, p. 134.
[4] Century Magazine, *loc. cit.*
[5] *Ibid.*

report to one of lower rank. To talk of a general officer being ordered to report to a captain or a guide "for position," or for any purpose whatever, is a glaring solecism which exposes any one who uses it. A little pains to recall the circumstances under which the orders were given would make their form appear very natural. Their purpose and substance could never have been doubtful till the remembrance of events a quarter of a century old had become dim.

The line of defences in front of Franklin had, quite early in the morning, taken the form of a fortified *enceinte*, the batteries with embrasures being connected by an infantry parapet. Both the enemy's and our own officers and men spoke of them as fortifications. An order to report to the commandant of a fortification "for position in the redan," for instance, carries with it the duty of serving there also, and not merely of finding a place. This usage was common at headquarters on the day in question, as is shown by the diary of Colonel Fullerton, Stanley's chief of staff. He notes at 1 P. M., that General Wagner is supposed to be moving his division "within the bridge-head constructed by General Cox around the town of Franklin."[1] With this somewhat formal idea of "the works," there was no danger that any officer ordered to report to me for position there would doubt his subordination. No one, in fact, did doubt it. It required twenty-five years to create that illusion.

The matter, then, is a very simple one. To meet a temporary exigency, General Schofield ordered certain detachments from the Fourth Corps to report to the commandant of the Twenty-third. When the exigency was past, the same power would order them

[1] O. R., xlv. part i. p. 149.

back. It could, of course, do so at any moment. To speak of this as removing Stanley from the command of his corps is nonsense. It is not pleasant to find the units of one's command scattered, but it often happens, and the loyal subordinate submits without a murmur. At the battle of Chickamauga one of the incidents was that the divisions of McCook's corps were sent to reinforce Thomas, till he was left with only Sheridan's. At the battle of Atlanta my division of five brigades (including one of dismounted cavalry) was scattered by Sherman's order, until I was left with only one, covering the extreme left of the army on the Decatur road. Nobody has any right to complain of such things.

Had Hood concluded to turn our position at Franklin instead of attacking in front, General Schofield would no doubt have withdrawn Kimball's and Wagner's divisions from the south side of the Harpeth, possibly sending Ruger also to report to Stanley, and leaving me alone with one division in my works. In the forenoon this seemed most probable. It was the fortune of war that it turned out otherwise.

A brief reference to the orders given to General Wagner will complete all that it is necessary to say of the troops on the line. The tenor of those orders need not be repeated.[1] If he had withdrawn from the front and come within my "bridge-head" in accordance with them, his whole division would have been, where Opdycke's brigade was, in reserve behind the line and in close support. General Schofield's official report tells us that such was his intention.[2] There is also no dispute that an order placing troops thus in reserve gives to the com-

[1] *Ante*, pp. 66, 71.   [2] See Appendix A.

mandant of the principal line the power to call upon them when needed. Opdycke's brigade was thus called upon,[1] and in the "nine points" approved by General Stanley the rightful subordination of that command to my orders is unquestioned. The disorganization of the other two brigades made it impracticable for them to act as reserves, and the scattered groups which rallied at the works fell under the command of the organized brigades in which they happened to be.

With this understanding of the situation, and of the application of military rules, what might happen on the line was provided for. General Schofield could leave the south side of the river and the village, and go to any central position north of it, from which he could be equally near to the cavalry corps, to Wood's division of the Fourth Corps, and to the Twenty-third Corps artillery, all of which were on the north bank of the Harpeth. He would thus be prepared for most efficient control of all his forces, whatever might turn out to be the plan adopted by his adversary, General Hood. General Stanley could properly accompany his superior to be ready promptly to take the immediate command of troops used to checkmate Hood in any other move than that of a direct and rash attack upon my works.

If I were not in the relation to the line in front of the town which I have described, the arrangement would have been radically faulty. General Stanley should in that case have remained in as close touch with the troops he claimed to command as I was with the Twenty-third Corps. We all knew beforehand what the event proved, that if anything went wrong on any part of that line, there would be no time to

[1] *Ante*, p. 95.

correct it by reporting it to the Truett house or even to Fort Granger, north of the river. It must be corrected instantly and on the spot. The fact, therefore, that General Schofield and General Stanley both moved their headquarters from the village over the river soon after noon helps to prove the complete understanding that the whole "bridge-head around the town of Franklin" was under my command. On this view, contingencies were provided for, and a responsible head was left on the line. On any other, they were not.

In the letters in the Century Magazine,[1] the suggestion appears that under the Articles of War a momentary authority might have devolved upon me by reason of portions of the Fourth Corps troops happening to do duty with mine, upon the line, whilst I was senior there, and that my right to command would thus be dependent upon military law, and not on the tenor of the written orders. I then replied that I should be entirely indifferent as to the source of power, were it not that the facts proved that everybody, at the time, recognized it as dependent upon the orders, as General Stanley himself had most explicitly done in the "nine points."

It may be interesting, however, to inquire briefly as to the application of the military rule. The 62d of the Articles of War[2] is this: "If upon marches, guards, or in quarters, different corps of the army shall happen to join or do duty together, the officer highest in rank of the line of the army, marine corps, or militia by commission, there on duty or in quarters, shall command the whole and give orders for what is needful to the service, unless otherwise specially directed by the President of the United

---

[1] *Ante*, p. 259.   [2] Army Regulations (1861), p. 508.

States, according to the nature of the case." Very good authority could be quoted for holding that this law meant by "different corps," those differing in function, as the engineers, ordnance corps, marine corps, or medical corps, and that it does not apply to different bodies of troops, like companies, regiments, brigades, divisions, etc. The "army corps," as everybody knows, was created long after the article quoted was enacted. The right to command in general is much more fundamental, and is found in the very first paragraph of the Regulations, which says that "all inferiors are required to obey strictly and to execute with alacrity and good faith the lawful orders of the superiors appointed over them."[1] As is usual in such matters, the difficulty arises when it becomes necessary to decide who is the "superior," has he been "appointed over" another, and what are "lawful orders."

Waiving all technical discussion of the meaning of the phrase "different corps" in the Article of War, it is very clear that its principle could only apply in the case of the absence of a common superior. The captain of Company A in a regiment does not acquire any command over the captain of Company F, who follows him on the march, or is next to him in line, by mere virtue of contiguity. If the regiment is under the command of a field officer, the captains are independent of each other, and strictly subordinate to the common superior present. If no field officer were present, the command of the whole regiment would devolve on the captain of Company A. The elementary principle applies to divisions or army corps as well as to companies.

[1] Army Regulations (1861), p. 9.

But if Companies A and F were separated from the regiment so that no field officer could exercise practical control, then, undoubtedly, in matters not explicitly provided for in previous orders, the two companies would become, by necessary implication, a new temporary organization for the common purpose, and by virtue of his seniority the captain of Company A would become the commander. Here, too, the rule for divisions or army corps is identically the same. In the larger organizations, however, it is more difficult to say when a common superior is present, and it is correspondingly difficult to say, as between two of his subordinates, when a case arises in which one could assume command over the other. As the grades of rank and command increase, the discretion of the subordinate increases in the mode of executing his own orders; but it does not increase in the matter of assuming authority over other equal organizations. On a field of battle, above all other places, it would be intolerable and disastrous for one corps or division commander to assume authority over another, by reason of contiguity, when there is a general in chief present in actual command. From top to bottom of the army organization this is a fundamental principle, under which Article 62 is only a grouping of particular cases. It applies to two corporal's guards as well as to two army corps.

In the battle of Franklin General Schofield was present, commanding three corps, two of infantry and one of cavalry. He had issued no orders putting either of his three principal subordinates under the command of another, and no one of them has ever pretended to have had such authority. General Stanley, in express terms, disclaimed it.[1] They

[1] O. R., xlv. part i. p. 118. See Appendix D.

each looked to General Schofield, and to him alone, for orders. No case had arisen, therefore, either under Article 62 or any other clause of the Regulations, for looking to anything else than the actual orders for the authority of his subordinates.

Our experience in the Atlanta campaign was a constant lesson as to the difficulties likely to arise from disputed authority when different bodies of troops came into casual connection. General Schofield found such a question arise between him and General Stanley, in the movement turning Atlanta, on the thirty-first day of August, and the authoritative decision of it was not announced till the 7th of November, after the opening of the new campaign.[1] It gave rise to a somewhat voluminous correspondence which was referred to Washington. Schofield himself was far too prudent an officer to leave such questions open among his own subordinates. From first to last in the campaigns of the year, he provided against dispute by explicit directions embodied in his orders.

As the Atlanta campaign opened, on the 1st of May, 1864, the Twenty-third Corps was at Charleston, Tenn., a little north of the Georgia line. The corps was ordered to march to Cleveland, Tenn., on the 3d, "in order of rank of division commanders." General Schofield intended to precede the troops to Chattanooga for consultation with General Sherman, and to rejoin the corps when it should reach Cleveland. Here was (if ever there was) a plain case under the principle of the 62d Article, and without special order I should have had, by virtue of my seniority, the command of the corps on the march. But General Schofield did not trust to

[1] O. R., xxxviii. part v. p. 734; xxxix. part iii. p. 685.

such implied authority. In the order itself, addressed to me, "he directs that you assume command of the whole, and conduct the march, and place the troops in position at or near Cleveland."[1]

Again, on the 3d of May, two divisions of the corps were ordered to advance to Red Clay in Georgia (now in the presence of the enemy); and "he directs you to take command of the two divisions and place them in position at Red Clay, and retain command until the arrival of the commanding general or till further orders."[2]

Passing to one of the latest incidents of the Franklin campaign, on the march from Pulaski to Columbia, General Schofield directed me to take the advance, and on November 21st wrote: "I desire you to move to Lynnville to-morrow morning. General Wagner will follow you. I will probably join you there before night. Reports indicate that Forrest is moving with Hood. He may, however, push out and strike the railroad to-morrow; if so, try to prevent him. In my absence take command of Wagner's division."[3]

In the face of such evidence of the custom not to trust to the understanding of the Regulations, can we be asked to believe that in preparing for the battle of Franklin, when Hood's formidable array showed the imminence of one of the deadliest and most persistent combats of the war, General Schofield intended to leave the question of command and responsibility on the line to the interpretation of the officers involved, and to their decision, without direction, in the excitement of the battle itself? He did no such thing. He gave written orders, which

[1] O. R., xxxviii. part iv. pp. 5, 6.   [2] Id., p. 23.
[3] O. R., xlv. part i. p. 974.

were unhesitatingly interpreted, both by myself and by every one of the officers ordered to report to me, as clearly fixing their subordination and my authority, as we have seen. In each case the official reports have demonstrated it, and the whole was in perfect accord with the purpose he had orally expressed at daybreak.

Having seen the organization of the line to meet the contingencies of battle or of withdrawal, and the transfer of headquarters of the army and of the Fourth Corps to the north side of the river, it will be natural to inquire whether the orders given and arrangements made were changed or affected by General Stanley's appearance at the front when the battle opened. To understand the matter clearly we need to learn just what General Stanley did. There has never been any pretence that General Schofield issued any new orders changing the command upon the line. Both in his letter to the Century, and elsewhere, General Stanley has contended only that the orders which have been already discussed had another meaning from that which I attributed to them. He was not ordered to come to the front or to assume any command. He contends that he had sufficient authority at all times during the day. The error of this having been shown, I shall inquire as to the military effect of his volunteered appearance on the line after settling as distinctly as possible what the facts actually were.

In his so called official report [1] General Stanley said: "From one o'clock until four in the evening the enemy's entire force was in sight, and forming for attack; yet in view of the strong position we

---

[1] The irregular paper, styled an official report, by General Stanley, Appendix D, will be examined presently.

held, and reasoning from the former course of the rebels during this campaign, nothing appeared so improbable as that they would assault. I felt so confident in this belief that I did not leave General Schofield's headquarters until the firing commenced."[1]

Colonel Wherry, General Schofield's senior aid and chief of staff, had been with me upon the line but a few minutes before, bringing final directions from the general in view of the expected engagement, and on his way back to headquarters at the Truett house on the Nashville road, was overtaken by one of my orderlies bearing the message that the enemy was about to attack. Hastening on across the bridge, within half a mile, he met Generals Schofield and Stanley riding toward the river. Schofield went into Fort Granger, and Stanley came to the front, where he and myself met on the Columbia Turnpike, near the Carter house.[2]

---

[1] O. R., xlv. part i. p. 115. Appendix D.

[2] *Ante*, pp. 98, 100. Colonel Wherry's letter to me, dated October 7, 1895, describing this incident, is worthy of fuller quotation, as giving interesting circumstances attending the opening of the battle, and showing the exigencies of staff service at such a time. In his statement his memory was assisted by original memoranda made at the time. Speaking of General Schofield's movements on that day, he proceeds:—

"Our first stop was early in the morning at the Carter house; then down at the river by the bridges. From there the General and his staff went to Dr. Clift's house with General Stanley about 10 or 10 30 A. M. We remained there, the staff going and coming as required by duty, until after dinner. About 2 P. M. Generals Schofield and Stanley removed their headquarters to the north bank of the river; to a house some distance out on the Nashville road. They gave personal attention to some details in respect to the placing of Wood's division as a reserve to cover the crossings of the Harpeth.

"About 3 P. M. I was sent with an order to the south bank, which having delivered, as I was suffering with a severe toothache from a tooth prepared for filling but not filled, I repaired to Dr. Clift's house again to get the doctor to draw the tooth. He was absent, and I

This meeting with General Stanley on the Columbia Turnpike, near the Carter house (our first and only meeting that day), has already been described,[1] and is one of the points on which he declared that we were exactly agreed. In my investigation in 1881 I found that General Opdycke took exception to the statement of General Stanley that his men were lying down when the latter reached them. He wrote to me (December 6th): "Before he [Stanley] came, I had received your request to have my brigade ready. It was *then* lying down by the stacks of muskets, having just got something to eat. The whole command must have been in order in a minute after your request came. I did not see Stanley nor know anything of his whereabouts till after the line was

waited for his return, leaving my horse in charge of my orderlies at the gate. Very shortly after 3.30 a messenger from you to General Schofield, with your information that the enemy was forming up and about to attack, came by the house which had been our headquarters up to 2 o'clock, and, recognizing my horse and orderlies, delivered his message to me. I mounted, and, taking the man along, rode across the bridge, and within half a mile — a few hundred yards my memorandum has it — I met Generals Schofield and Stanley riding toward the river. They had received from some other source the information I was conveying. I turned and rode with them. At the bridge General Stanley crossed and went to his troops at the Carter house, and General Schofield rode into Fort Granger, which, as you know, was very close to the railroad bridge and was the prolongation of our left, only separated from the line of battle on that flank by the narrow stream. It overlooked the front nearly to the Columbia pike at the Carter house, and was where communication with the reserves and line of battle could best be had. General Schofield's headquarters were there until after the battle, when we went to the same house out on the Nashville pike about a mile, where was General Stanley, who had been wounded, and his staff. There we remained until after midnight, when the head of column came along from the river and we rode to Brentwood hills."

Colonel Wherry was Brevet Brigadier General at the close of the war, and is now (1896) Lieut. Colonel Second Infantry, U. S. A., and Brevet Colonel.

[1] *Ante*, pp. 98, 100.

restored, when he rode up to me and showed me that he was wounded. He immediately rode away."

The omission to notice General Stanley at the instant of his arrival is not remarkable in the intense preoccupation in his own work at such a moment. The order of events is important as showing that under my direction he had already begun preparation for the movement forward.

The same letter bears directly upon the question next in order, How long did General Stanley remain upon the line? Immediately after the quotation last made, it says: "I have always insisted that his report did you a grave injustice. It simply represents you as gallantly assisting *him*, when the fact was that you were commanding the whole line of battle, and he aided you, I suppose about five minutes, and then left, wounded." In his published account of the battle, Opdycke stated with equal distinctness that, "soon after coming under fire, Stanley's horse was shot under him, and he himself wounded and so compelled to leave the field."[1]

There is no room for dispute as to the time and circumstances of General Stanley's receiving his wound. In the "report" he himself says it was "just after the retaking of the line by our troops."[2] In the eighth of the nine points to which he agreed it was said to be "when Opdycke reached the parapet," and we "were trying to rally the fugitives immediately in rear of the line."[3] All who were personally witnesses of what occurred corroborate Opdycke's statement as to the wounding in that first mêlée, and Stanley's immediate retirement for surgical treatment, as I urged upon him.

[1] New York Times, September 10, 1882.
[2] Appendix D.   [3] *Ante*, p. 262.

Major Dow, my Inspector General, was himself in the group at the turnpike, rallying the men as the reserves went forward. He says: "Generals Stanley and Cox were on the pike with them" (Opdycke's troops). "General Stanley was wounded in the first mêlée, and his horse shot also. I saw his wound in the neck, and saw him remounted on one of General Cox's horses. . . . Stanley exercised no command while there, but rallied the men."[1]

Captain Cox, Adjutant General, says: "General Stanley rode up to the front at this point; but he had scarcely done so (and before he had time to view the field, or give or suggest an order) when he was wounded and his horse shot. Captain Tracy then dismounted from your dun horse, which he had been riding, and gave it to General Stanley, who immediately mounted, and rode off through the village. This was the only time I saw General Stanley that day, until I went over the river in the evening to see General Schofield at his headquarters."[2]

Captain L. T. Scofield, Topographical Engineer, also speaking as an eyewitness, says in his paper on the campaign already quoted: "General Stanley was also there, showing great gallantry in encouraging the troops, but was wounded before he had been on the field ten minutes, and retired."[3]

General Reilly, under date of February 21, 1876, wrote me: "I did not receive orders that day, as you are aware, from any other than yourself, and supposed, as I do yet, that you were in chief command on south side of river. General Stanley, although wounded in rear of line of Third Division (Twenty-third Army Corps), did not give or assume any command."

[1] Appendix E.   [2] Appendix F.
[3] Ohio Loyal Legion papers, vol. ii. p. 137.

The earliest formal history of the Army of the Cumberland was that by Van Horne, who, though not present at this battle, had long been connected with General Thomas's staff, and had full access to all his papers at the time he wrote. He may therefore be considered as giving the general understanding at Thomas's headquarters when he said that Stanley "was pierced in the neck by a bullet and was compelled to leave the field." [1]

There are also statements of officers whose positions at and near the centre gave them the opportunity of knowing what general officers were or were not there after the first restoration of the line; and though negative in form, they are hardly less conclusive in fact than the testimony of those who saw General Stanley go to the rear at that early stage of the battle.

General Casement, whose brigade was next east of the cotton-gin, in a letter to me of October 21, 1889, (from which I have already quoted,[2]) after telling of his being "along the line several times to the Carter house" during the battle, says of General Stanley that he did not see him, and that "he was never on my portion of the line, either while in course of construction, after it was finished, or during the progress of the battle." He mentions also coming to me, near the Carter house, just about dark, and adds: "I know that there was no officer of superior rank to yourself on the field . . . I never heard of such an officer till long after the war was over." I had put to him the direct question,

[1] Van Horne, "Army of the Cumberland" (1875), vol. ii. p. 201. This book had been published seven years when my volume, "Franklin and Nashville," appeared.
[2] *Ante*, p. 146, note.

"From all you saw of my position and action at the time, and the conduct of others in recognizing my orders, what did you as an officer understand my relations to the line to be?" His answer was, "I understood you to be in command of the line, and so understand now; and it is too late to change the facts."

The 73d Illinois was one of the regiments of Opdycke's brigade of the Fourth Corps, and its officers were therefore General Stanley's subordinates. The regiment published its own history, and Captain E. J. Ingersoll was one of the contributors to it. He says of his own movements immediately after the restoration of the line: "I passed from the Carter house to the cotton-gin, then returned to first piece of artillery in rear of Carter house. . . . Met General Cox on the pike in front of the Carter house about 5 P.M., and again about 10 P.M. — the only general officer I saw on the ground."[1]

Captain Sexton, in command of the 72d Illinois, in Strickland's brigade, in his communication from which I have already quoted,[2] says: "The only superior officer I saw after the battle opened was General Cox."

When it is remembered that the gist of General Stanley's contention is that he remained throughout the battle on the line, and at the centre, which was the critical point, the appositeness of these quotations is plain.

Our next step is to find where General Stanley received the surgical attention which he went to seek when he left me on the line at the Carter house.[3] I have already quoted from a letter of Surgeon Hill[4]

[1] History of the 73d Illinois Regiment, p. 642.
[2] *Ante*, p. 254.   [3] *Ante*, p. 101.   [4] *Ante*, p. 95.

(of Kimball's division), who stood near General Schofield on the parapet of Fort Granger, and saw the opening of the fight. He then went to his field hospital, which he says was "some distance to the right of the fort (looking toward the south), a little to its rear, and north of the river — also a little west of the Nashville pike." This description accurately locates the field hospital a short distance northwest of the place where the temporary road from the wagon bridge over the Harpeth joins the Nashville Turnpike on the north side of the river.[1]

To this hospital General Stanley went, still in wrathful excitement from traversing the crowds of disorganized men and stragglers in the village, aggravated by the painful smarting of his wound. He at first mistook the group at the hospital for a crowd of skulkers, and it took, Dr. Hill says, a little time to get him to dismount. "An examination revealed a flesh wound running transversely across the nape of the neck, probably three inches long and equal in depth to half the diameter of the ball. The wound stung furiously. . . . After quieting down, the wound was dressed, an opiate administered, and he was put in an ambulance, which moved off up the Nashville road. . . . He must have met me," the surgeon concludes, "within half an hour of the receipt of his injury, and was not to exceed a half hour in my hands."

As soon as the repulse of the enemy's first assaults gave assurance that we should be able to hold our lines against him, and probably immediately upon the receipt of the order to withdraw at midnight,[2]

[1] Surgeon Hill was also kind enough to locate the place upon a sketch traced from one of the published maps.

[2] *Ante*, p. 169.

I sent my Adjutant General to General Schofield's headquarters to express my strong confidence that we could await the expected reinforcements where we were. This was very soon after dark. He says of it: "At General Schofield's headquarters, besides his usual staff, I saw General Stanley and several officers."[1] A little later, on a call to consult General Schofield's Inspector General in reference to the placing of pickets and skirmish line when we should retire, I sent Major Dow to the same headquarters. He, too, found Generals Schofield and Stanley together at the Truett house.[2]

General James H. Wilson, the cavalry commander, in a paper upon the campaign published in the Century War Book,[3] also mentions meeting Stanley at General Schofield's headquarters. He says that "immediately after the close of the cavalry battle . . . I rode to General Schofield's headquarters. . . . Schofield and Stanley, the latter severely wounded, were together, discussing the events of the day."[4] The dispatches which passed help to fix the time pretty accurately. At five o'clock (just as the sun was setting) General Wilson had been able to announce the Confederate cavalry defeated and driven back across the Harpeth. It was a half hour's ride from his headquarters to General Schofield's, and the latter acknowledged the receipt of the welcome dispatch at half past five.[5] Waiting

---

[1] Appendix F.   [2] *Ante*, p. 170, and Appendix E.
[3] Battles and Leaders, vol. iv. p. 466.
[4] In the paper referred to, General Wilson speaks of the headquarters as being in Fort Granger. It is agreed on all hands that they were moved from the fort to the Truett house when it became too dark to see the lines from that point. The slip of memory as to the place is corrected by Colonel Fullerton's diary.
[5] O. R., xlv. part i. p. 1179.

only to make sure that his front was really quiet, General Wilson tells us he "immediately" rode in person to see Schofield, as he had been asked to do in an earlier message.[1] In the first darkness of the evening (for it was dark within half an hour after sunset),[2] General Wilson found Generals Schofield and Stanley together, the wound of the latter having been already dressed, and he sufficiently comfortable to join in the discussion of the events of the day. After this visit General Wilson returned to his headquarters at the cross-roads two miles and a half east of Franklin, where he was again established in time to write instructions to General Hammond, his subordinate, dated at 8.45 P.M.[3] The mention by Colonel Wherry of General Stanley's presence at headquarters when General Schofield reached the Truett house, has already been noted.[4]

After General Stanley's arrival at the Truett house, before seven o'clock in the evening, Colonel Fullerton, his chief of staff, was prepared to resume the annotation of events in his diary, and had entered an outline of what had occurred between four and seven o'clock.[5] A series of entries runs through the evening. At seven it is noted that "the enemy has been steadily fighting up to this hour since 4 P.M." At half past seven the visit of General Wilson is mentioned. At eight the preparations for withdrawing from the lines are spoken of, but it is said to be "very doubtful whether these troops can be withdrawn, as they are very close to the enemy,

---

[1] O. R., xlv. part i. p. 1178.   [3] O. R., xlv. part i. p. 1184.
[2] *Ante*, p. 147, note.   [4] *Ante*, p. 281, note.
[5] The hours at which entries are made in such diaries must be taken with due allowance. The outline referred to is dated 4 P. M., but its contents show that it must have been written after Colonel Fullerton's return with General Stanley to the headquarters.

with whom they are keeping up a continual skirmish fire," and the Confederates still making frequent though feeble assaults. At nine the reports are that "the enemy is relieving his troops, which have been engaged, and is substituting others for them." At ten the fire in the town is noted, supposed to be started "by the enemy in order to show them any attempt we might make to retreat," and which "with difficulty was put out by midnight." The retirement of the troops then began.[1]

It is not often that the movements of a single officer can be so consecutively accounted for as in this instance: the circumstances also become vividly clear. Through the long evening it is very plain that everybody at General Schofield's headquarters appreciated the fighting going on upon the line, and

---

[1] O. R., xlv. part i. pp. 150, 151. If further evidence of the prolonged night engagement were needed, one cannot go amiss in referring to any of the official reports on either side.

General Hood, in his official report, said that "the struggle lasted till near midnight." O. R., xlv. part i. p. 653. In another place that "the struggle continued with more or less violence till nine o'clock in the evening, and then much desultory fighting through most of the night." Advance and Retreat, p. 295.

General Schofield reports that "the enemy assaulted persistently and continuously with his whole force from about 3.30 P. M. until after dark, and made numerous intermittent attacks at a few points until about 10 P. M." O. R., xlv. part i. p. 342, and Appendix A of this volume.

My own report states that the assaults on Ruger's division and Opdycke's brigade "were obstinately repeated till nightfall, and even as late as nine o'clock attacks were made, which were, however, easily repulsed, and the enemy withdrew the remnants of his shattered lines to the position occupied at the opening of the battle by Wagner's division, in advance of our lines about eight hundred yards. Alarms occurred frequently until eleven o'clock, and frequently caused a general musketry fire on both sides from our centre toward the right, but I found no evidence that any real attack was made at so late an hour." O. R., xlv. part i. p. 354, and Appendix B of this volume; also chap. xii., *ante*.

the danger that at any moment Hood's efforts would resume the desperate character of the earlier assaults. The delicacy of the operation of withdrawing was also evidently discussed and commented upon. If the Fourth Corps troops in the line were not with the men of the Twenty-third Corps under a common head, empowered to meet such emergencies as might momently arise, it is too plain for argument that a commander for the Fourth Corps ought to have been by my side near the Carter house, mutually to assist in instant decision and in action. The Fourth Corps staff diary, like every other contemporaneous report, emphatically contradicts the assumption that the battle was over when General Stanley left the front, and the distance between the Carter house and the Truett house was altogether too great for communication by messenger before acting in emergencies that might arise.

The simple sequence of facts destroys all such fictitious theories of the situation. General Stanley left the lines, wounded in the first mêlée. My task of inspection immediately followed.[1] Orders were sent to Kimball's division of the Fourth Corps for the reinforcement at the centre, and Colonel McDanald marched his regiment to its assigned position in Ruger's line before the setting sun marked that five o'clock was reached, and two hours before the hour at which Colonel Fullerton entered in his diary, "The enemy has been steadily fighting since four o'clock."

The Twenty-third Corps headquarters on the Columbia Turnpike, near the Carter house, was the only place with which communication was kept up with reference to the conduct of the battle at the

[1] *Ante*, p. 144.

works; and the messages sent and received through both General Schofield's staff officers and my own fully implied my complete reponsibility for affairs there. On this point the statement of Colonel Opdycke, which I have quoted,[1] is supported by that of Captain Scofield, who says: "I knew early in the day that you had been placed in command of the troops in the fortifications south of the Harpeth River. I did not see the written order assigning you to command, but during the day and night I was present when officers from General Schofield's headquarters brought such instructions and received such information as would only come to and from one in active command."[2]

Among the staff officers who were indefatigable in duty that day and night, none was more active than Colonel Wherry, General Schofield's chief of staff, and none knew better the relations of officers on that field. It has more than ordinary conclusiveness, then, when he, before the publication of my volume on that campaign, wrote: "Not only has it been attempted to withhold from General Schofield the measure of honor due him as commander of the combined forces in the field, including Franklin, but your part as commanding the line at Franklin has been usurped by others."[3]

General Stanley himself becomes a witness to the truth of the situation thus described, when, remembering that Wagner was reorganizing two of his

[1] *Ante*, p. 282.
[2] Letter of November 3, 1889. See also statement of Surgeon Frink, *ante*, p. 40, note.
[3] Letter of June 21, 1881. I do not understand that General Wherry intended to express a judgment as to any theoretic right to command during the time General Stanley was actually present on the line.

brigades in the town, we find him in the last of the "nine points" assenting unequivocally to the fact that "Opdycke and the artillery continued to act under my orders till we left the lines at midnight."[1] What Kimball, his other division commander, did, is told by the official report of that division.[2]

The inquiry propounded at the beginning of this chapter as to the military effect of Stanley's brief appearance at the front may be answered by saying that the orders under which troops are detached continue in force until they are revoked by the authority which issued them, or the duty is completely performed. "On the return of a detachment the commander reports to the headquarters from which he received his orders," and no officer can put himself on duty without orders from competent authority.[3]

Portions of the Fourth Corps having been temporarily detached, and ordered to report to another officer, General Stanley could not, at his own will, resume command of them. However gallant the motive, he appeared at the lines simply as a volunteer. He did not in fact issue any orders, and as he soon retired, wounded, no question arose as to his status or his authority.

How then can we account for his intemperate assaults upon the veracity of statements which are carefully within the limits of the truth so abundantly sustained as we have seen? We must look for the answer in the examination of the document which he calls his "official report."

[1] *Ante*, p. 262.
[2] *Ante*, p. 263.
[3] Army Regulations of 1861, §§ 8, 655.

After the battle of Nashville (December 15, 16, 1864), General Thomas called upon his subordinates for official reports of the preceding campaign. These were naturally divided into two parts: first, the campaign in the field, including the battle of Franklin, when General Schofield was in command of the active army, consisting of the Fourth and Twenty-third Corps and the Cavalry Corps; and second, the subsequent period, including the battle of Nashville, when General Thomas was personally in command. For the former period the reports should all pass through General Schofield's hands, his own report being a general one for the whole army in the field. For the latter period the corps commanders would collect their subordinates' reports and forward them with their own to General Thomas direct.

In the regular report of the Fourth Corps for the latter period, General Wood, the senior division commander,[1] says: "Maj. Gen. D. S. Stanley, having been wounded in the conflict at Franklin, on the 30th of November, and having received a leave of absence on account of his wound, relinquished, and I assumed command of the corps on the 2d of December."[2] Being thus in command, General Wood, on the 10th of January, 1865, made the regular official report of the Fourth Corps for the first period of the campaign. He says that he did it "in pursuance with orders received from the headquarters Department of the Cumberland," and that "the duty devolved upon" him, in consequence of the relinquishment of command by General Stanley.[3] The report was a full and detailed one, and no one

[1] General Thomas J. Wood, now Brevet Major General, U. S. A., retired.
[2] O. R., xlv. part i. p. 126.   [3] *Id.*, pp. 119, 126. Appendix C.

will question General Wood's title to equal ability and experience with General Stanley. With this regular official report of the Fourth Corps went up the subordinate reports. On the same date I made the report for the Twenty-third Corps.[1] General Schofield's report for the active army in the field was made on December 31st.[2] Upon the receipt of these reports, General Thomas prepared his own, which is dated January 20th, and the whole, with the accompanying papers, went to Washington.[3]

General Stanley's absence from the army continued through the crisis of the campaign, the battle of Nashville, the pursuit of Hood, and until the latter part of February, when he resumed command of the Fourth Corps at Huntsville, Ala. In January, General Schofield and the Twenty-third Corps were transferred to the Atlantic coast, and our connection with the Army of the Cumberland was finally severed. After all this, on the 25th of February, 1865, General Stanley writes what he styles a "Report of the services of the Fourth Army Corps from the time of separating from General Sherman's Army at Gaylesville, Ala., to and including the Battle of Franklin, Tenn."[4]

This paper wholly ignores the regular official report of the corps made in ordinary course of duty by General Wood, and traverses the identical ground, as if no such report had been made. General Wood's report is in the usual impersonal style of military reports, stating what was done by the corps and its parts. General Stanley's is throughout much more

[1] O. R., xlv. part i. p. 349. Appendix B.
[2] *Id.*, pp. 339–347. Appendix A.
[3] *Id.*, pp. 32–46.
[4] *Id.*, xxxix. part i. p. 907, and xlv. part i. p. 112. Appendix D.

in the form of a personal narrative, emphasizing his personal part and his opinions. The matter, however, is in substance the same until the affair at Spring Hill (November 29) and the battle of Franklin (November 30) are reached. Here the *raison d'être* of the new paper becomes plain. The personal relations of the author of it to the battle is the theme. It is not, however, with the style or taste of the document that we have now to do, but with its official character and its value as a narrative.

It was wholly irregular. The official report of the corps had been regularly made in accordance with the rules of the service. If anything in General Wood's report had been unjust or erroneous, or if anything really important was omitted, it was General Stanley's privilege, on its coming to his knowledge (as it would have been of any other officer in the service), to address a letter to the Adjutant General pointing out the errors. This would lead to official inquiry in an established method. The character of the new matter in the so called report could in no way be so well shown as by culling out all that is additional to General Wood's report, and embodying it in the form of such a letter to headquarters. It consists, in the main, of the personal experience of General Stanley on the field. The description of the battle would give the reader no suspicion of the actual relations of the troops of the two corps. One would not guess that Opdycke's brigade and the batteries were the only organized bodies of the Fourth Corps in the lines near the centre; that Ruger's whole division of the Twenty-third Corps was in line between Opdycke's brigade and Kimball's division; and that Reilly's division of the Twenty-third Corps extended on the left from

Opdycke's brigade to the river. We get no assistance in learning why General Wagner and his two brigade commanders were not on the line, or where and how the two brigades were reorganized. The things which would enable the commander of the army to get definite knowledge of the battle through the relations of the several bodies of troops to each other and to the whole are conspicuously absent.

General Stanley's appearance on the field and what the soldiers said are dramatically told, but no mention is made of his visit to Surgeon Hill's field-hospital north of the river, or his spending the rest of the evening at the Truett house. On the other hand, after saying that his wound "did not prevent his keeping the field," the continuous narrative presents him in undiminished closeness of contact with the fighting lines in each successive incident until "at midnight the withdrawal was made successfully." That we may have no doubt of his meaning, he quoted the passage, in his newspaper letter, as proof of the falsity of the assertion that he "was reluctantly persuaded to return to his quarters for surgical help."[1]

But two or three sentences in the document are worthy of more careful examination. He says: "Just after the retaking of the line by our troops, as I was passing toward the left to General Cox's position, my horse was killed, and no sooner had I regained my feet than I received a musket ball through the back of my neck. My wound, however, did not prevent my keeping the field, and General Cox kindly furnished me a remount."[2] A couple of pages farther on, at the end of the lively narration, he adds: "Although Brig. Gen. J. D. Cox was not in my

[1] March to the Sea, p. 90.   [2] O. R., xlv. part i. p. 116.

command, he was my close neighbor in the battle of Franklin, and I take this opportunity to express to him my thanks for his gallant help at that time." [1]

I have already referred to the first of these quotations as fixing the time at which he was hurt, in substantial accord with my own recollection submitted to him in 1881,[2] and said by him to be exactly correct. The next statement, however, demands notice. What can be meant by "passing to the left to General Cox's position"? We were in the highway of the Columbia Turnpike, in front of the Carter house, where my headquarters had been since daybreak, where they remained till midnight, and where they are marked on Captain Twining's official map. That we went forward together, were together rallying the troops, when he was hit, that he asked me to examine his wound, and that I urged him to go for surgical attention, General Stanley perfectly remembered in 1881.[3] As to position, we were not only at my headquarters, but were at the centre of the Twenty-third Corps line; he could not possibly be more exactly at my military position or my bodily station. Yet the reader of this "report" would certainly understand that here was something like a passing from one corps position to another, off at the left!

As to his "keeping the field," I need not repeat the statements of the eyewitnesses (including myself), who saw him ride away toward the village, and who know that he did not return,[4] or of those by whom his whereabouts, from half an hour after he was hurt till midnight, are consecutively given. The facts were known to such a cloud of witnesses,

[1] O. R., xlv. part i. p. 118.
[2] *Ante*, p. 262.
[3] See map, *ante*, p. 45; also p. 262.
[4] *Ante*, p. 282, *et seq.*

including nearly every general and staff officer in the army, that one wonders at the rashness which could raise any question about it, until the so called report is read, when the assumption of heat in sustaining its assertions becomes intelligible. With the final artistic touch, thanking me for the aid I had incidentally given him in his arduous and continuous labors, the sketch is complete. Anything inconsistent with it must be stormed down as falsification. As he had probably forgotten the sweeping assertions of this report when, in 1881, appealing to his actual recollection,[1] he pronounced the "nine points" exactly correct, so in 1889 we may assume that the emphatic indorsement of the "nine points" was forgotten, and the thick-and-thin reiteration of the "report" was in order.

Except for such forgetfulness, the remembrance that he had distinctly agreed, as the basis for a historical publication, upon the fact that Opdycke's brigade and the batteries "continued to act under my orders till we left the lines at midnight,"[2] would have suggested the inquiry what need there was of his remaining at the lines or returning to them after the first repulse of the enemy at the centre. Those organizations were the only Fourth Corps troops in line there. The disorganized groups or individuals scattered among the rest were not in need of his presence. Wagner's two other brigades were reorganizing near the river, as we have seen.[3] Kimball's division was three quarters of a mile away on the extreme right, had easily repulsed the rebel cavalry, and was not again attacked. It is not alleged that General Stanley went there. What then was he in active command of, at the front, after he was

[1] *Ante*, p. 260, et seq. [2] *Ante*, p. 262. [3] *Ante*, p. 146.

wounded, that he so hotly treats as calumny the statement that he did the sensible thing of going to have his hurt attended to? He himself being the witness, the answer must be, of nobody and of nothing. Opdycke and the batteries being under my orders, the line of my command is continuous and complete from the Harpeth on the left to Carter's Creek Turnpike on the right. Kimball neither needed nor had any orders as to the right flank beyond Ruger, except to send the reinforcement to the centre, and that order went from me, as he officially reported.[1] We search the Official Records in vain for evidence of any order issued by General Stanley in the conduct of the battle, or any recognition of his presence on the line after he was wounded. It was, then, the fatal necessity of sustaining that extraordinary "report" that led to General Stanley's persistent and violent attacks upon other people's veracity.

To complete the consideration of the subject, it is only necessary to examine the contemporaneous records of the engagement which I myself made, so that the view I took of it at the time may be compared with the historical account written in 1882. As a memorandum of official action, I kept a brief journal, which, like similar headquarters diaries, has been published with the Official Records.[2] In this, under date of November 30th, the day of the battle, I entered the following: "Reached Franklin before day, having marched twenty miles during the night, passing the rest of the army on the way. No means for crossing Harpeth River, and I am ordered to take both divisions Twenty-third Corps, and hold a line above the town till the trains and the rest of

[1] *Ante*, pp. 263, 264.   [2] O. R., xlv. part i. p. 356, *et seq.*

the army are over. Enemy follows close, and two divisions of Fourth Corps — Kimball's and Wagner's — are ordered to report to me. Enemy assault at 3.30 P.M. Gain some temporary advantage in the centre at first, but are soon repulsed with terrible slaughter. We held the lines till midnight."

On the next day, December 1st, I noted in the journal the march to Nashville, and a brief outline of the results of the battle, concluding with this: "Whole loss of the army about 2,000, of which most was in Wagner's division, which was driven in from the front in confusion."[1]

In my preliminary report,[2] dated December 2d, I stated the duty intrusted to me in substantially the same form as in my journal, and that Kimball's and Wagner's divisions of the Fourth Corps reported to me by order of General Schofield.

On December 3d I recommended Colonel Opdycke for promotion, giving as a reason for doing so that his heroic service was whilst under my orders, "as I was commanding the line at that time."[3]

It is thus clearly shown that my view of my relations to the troops and to the line is not one that has been born of long brooding over the stirring affairs of that campaign, but was that which I held at the time, and which is sustained by the mass of evidence both official and private. But in my reports, sincerely sympathizing with an officer who was wounded in the performance of a gallant act in circumstances that gave him no official position, I understated my own authority, and attributed to him a command which he did not exercise. The reassertion of the truth in regard to this matter has

[1] O. R., xlv. part i. pp. 358, 359.   [2] *Id.*, p. 348.
[3] *Id.*, p. 409, and *ante*, p. 228.

not been of my seeking. My historical account of
the battle was entirely consistent with the "nine
points" which General Stanley had unqualifiedly
assented to for the express purpose of such a publication.[1] He has himself forced the comparison between the facts and his so-called report of February
25, 1865. He has also brought out the truth that
in my official report I claimed much less than the
reports of the officers of his own corps show that I
was entitled to.

After I had thus voluntarily diminished the statement of the extent of my authority on the actual
line, I was of course estopped (to use General Schofield's word) from calling public attention to the
truth until General Stanley's so called report came
to light. It was a number of years after the war,
when I first heard of that document. It completely
removed any cause for delicacy in thoroughly reviewing the facts, and, in the preparation to write the
history of the campaign, I was considering this
matter, when General Stanley himself took the initiative by addressing me the letter already mentioned.[2]
If he had then challenged the accuracy of the "nine
points" I submitted to him in reply, the discussion
of the subject would have been fully opened. As,
however, he unqualifiedly assented to their exact
correctness, I was glad to find an accepted outline of
a narrative in which controversy could be avoided.
I informed General Schofield, General Wherry, his
chief of staff, and General Opdycke, and they all
expressed their satisfaction that such an outline had
been agreed upon.[3] The volume was published, and

---

[1] *Ante*, p. 262, and March to the Sea, chap. v.   [2] *Ante*, p. 262.
[3] All this correspondence took place in 1881, and the letters are in
my private files. General Schofield's was dated December 5, and said:

nearly seven years passed, when the literary peace was wantonly broken in the manner which has been sufficiently stated.

In the light of the evidence which has been given, the candid reader can judge who, if anybody, has grasped at honors which were unearned. The means are offered of testing by the Official Records the vanity or the modesty, the selfish or the generous motives, of the reports of the campaign. The lesson is one which I even think may be useful to the careful student of our war history as showing how the abundance of the official documents and private material gives unexpected clues to the truth, and helps in the exposure of false assumptions.

A reference to my method in preparing both my little volumes of war history, the "Atlanta" as well as the "March to the Sea," will close what I have to say. Both General Sherman and General Schofield scrupulously refrained from pressing upon me their own views of their campaigns, and strictly limited their assistance to furnishing me copies of papers, maps, and documents. They concurred with me that it was not best that they should see any part of my work till it should appear in print. I was happy to find, when it was published, that they both regarded the narrative as authentic and authoritative. After the volume on Franklin and Nashville appeared, General Schofield kindly gave it a careful reading, and noted even minute errors for my assistance in revising for a new edition. In the account of the battle of Franklin he noted none, and suggested no

---

"It is very gratifying to me that you and Stanley have, upon a comparison of views, found yourselves in substantial accord in respect to the relations of your two commands at Franklin. I believe your way is now clear to a just and impartial account of that battle."

changes. It was with a satisfaction which was not unnatural that I had found him saying of that campaign in general: "No doubt you must be prepared to meet unjust assaults upon your history; but you are fully fortified by the evidence and by sound military principles. You may well stand fast on the record, as you have stated it, and I shall take great pleasure, if there be any need, in defending the truth as you have written it."[1] In his letter written after the more careful reading,[2] the tone is the same, and it is in it that he spoke of my being possibly "estopped" from a more critical treatment of these events[3] by "the generous treatment already given" to others. Such commendation seemed to me to set the seal of candor and fairness upon my work.

[1] Letter of September 13, 1882.
[2] November 29, 1882.
[3] See *ante*, p. 265.

APPENDIX A

GENERAL SCHOFIELD'S REPORT

Dated December 7, 1864

(Part relating to the Battle of Franklin[1])

I ARRIVED at Franklin with the head of column a little before daylight on the 30th, and found no wagon bridge for crossing the river, and the fords in very bad condition. I caused the railroad bridge to be prepared for crossing wagons, and had a foot bridge built for infantry, which fortunately also proved available for wagons, and used the fords as much as possible. I hoped, in spite of the difficulties, to get all my material, including the public property, and a large wagon train at Franklin, across the river, and move the army over before the enemy could get up force enough to attack me. But I put the troops in position as they arrived on the south side, the Twenty-third Corps on the left and centre, covering the Columbia and Lewisburg pikes, and General Kimball's division of the Fourth Corps on the right, both flanks resting on the river. Two brigades of General Wagner's division were left in front to retard the enemy's advance, and General Wood's division, with some artillery, was moved to the north bank of the river to cover the flanks should the enemy attempt to cross above or below.

The enemy followed close after our rear guard, brought up and deployed two full corps with astonishing celerity,

[1] O. R., xlv. part i. pp. 342-344.

and moved rapidly forward to the attack. Our outposts, imprudently brave, held their ground too long, and hence were compelled to come in at a run. In passing over the parapet they carried with them the troops of the line for a short space, and thus permitted a few hundred of the enemy to get in. But the reserves near by instantly sprang forward, regaining the parapet, and captured those of the enemy who had passed it. The enemy assaulted persistently and continuously with his whole force from about 3.30 P. M. until after dark, and made numerous intermittent attacks at a few points until about 10 P. M. He was splendidly repulsed along the whole line of attack. The enemy attacked on a front of about two miles, extending from our left to our right centre, General Kimball's left brigade. Our two right brigades were only slightly engaged. I believe the enemy's loss in killed and wounded cannot have been less than 5,000, and may have been much greater. We captured 702 prisoners and 33 stand of colors.

Our loss, as officially reported, is as follows.[1]

I am not able at this time to give fully the names of the killed and wounded officers. Among the latter is Major General Stanley, commanding the Fourth Corps, who was severely wounded in the neck while gallantly urging forward his troops to regain the portion of our line which had been lost. General Stanley is deserving of special commendation, and has my hearty thanks for his cordial support and wise counsel throughout the short but eventful campaign. Brigadier General J. D. Cox, commanding temporarily the Twenty-third Corps, deserves a very large share of credit for the brilliant victory at Franklin. The troops were placed in position and intrenched under his immediate direction, and the greater portion of the line engaged was under his command during the battle. I recommend General Cox to

[1] Tabular statement omitted. It will be found in chap. xvi., *ante*. For the final statement of the enemey's losses, see the closing paragraph of this report.

Appendix A 307

the special consideration of the Government. Brigadier General Ruger, commanding Second Division, Twenty-third Corps, held the weakest portion of our line, and that upon which the enemy's assaults were most persistent. He is entitled to very great credit. Brigadier General Reilly, commanding (temporarily) the Third Division, Twenty-third Corps, maintained his lines with perfect firmness, and captured twenty battle flags along his parapet. I am also under great obligations to the division commanders of the Fourth Army Corps, Brigadier Generals Wood, Wagner, and Kimball, for the admirable manner in which they discharged every duty, and cannot refrain from expressing my high commendation, though in advance of the official report of their immediate commander. Colonel Emerson Opdycke, commanding First Brigade, Second Division, Fourth Army Corps, the reserve which recaptured the lost portion of our line, is spoken of by Generals Stanley and Cox as having displayed on that occasion the highest qualities of a commander. I cordially indorse their recommendation. For other special instances of gallantry and good conduct I must refer to subordinate reports.

On my arrival at Franklin I gained the first information from General Wilson since the enemy commenced his advance from Duck River. I learned that he had been driven back and had crossed the Harpeth above Franklin on the preceding day, leaving my left and rear entirely open to the enemy's cavalry. By my direction he sent General Hatch's division forward again, on the Lewisburg pike, to hold Forrest in check until my trains and troops could reach Franklin. This was successfully done, and General Hatch then retired before a superior force, and recrossed the river, connecting with my infantry pickets on the north bank early in the afternoon. A short time before the infantry attack commenced, the enemy's cavalry forced a crossing about three miles above Franklin, and drove back our cavalry, for a time seriously threatening our trains, which were accumulating on the north bank, and moving toward Nashville. I sent General Wilson orders, which he had, however, anticipated, to drive the enemy back at all hazards, and moved a brigade of General Wood's division to support him if necessary. At the moment of the first decisive repulse of the enemy's infantry I received the most gratifying intelligence that General Wilson had driven the rebel cavalry back across the river. This rendered my immediate left and rear secure for the time being. Previous to the battle of the 30th I had ordered all trains except ammunition and hospital wagons to Nashville, preparatory to falling back from Franklin when it should become necessary, which I expected on the following day. The enemy, having nearly double my force of infantry and quite double my cavalry, could easily turn any position I might take and seriously endanger my rear.

Only one division of the enemy's cavalry had been engaged with General Wilson during the 30th. The remaining three divisions were free to strike my line of communications, which they could easily do about Brentwood by daylight the next morning. My experience on the 29th had shown how utterly inferior in force my cavalry was to that of the enemy, and that even my immediate flank and rear were insecure, while my communication with Nashville was entirely without protection. I could not even rely upon getting up the ammunition necessary for another battle. To remain longer at Franklin was to seriously hazard the loss of my army, by giving the enemy another chance to cut me off from reinforcements, which he had made three desperate though futile attempts to accomplish. I had detained the enemy long enough to enable you to concentrate your scattered troops at Nashville, and had succeeded in inflicting upon him very heavy losses, which was the primary object. I had found it impossible to detain him long enough to get reinforcements at Franklin. Only a small portion of the infantry and none of the cavalry could reach me in time to be of any use in the battle, which must have been

Appendix A

fought on the 1st of December. For these reasons, after consulting with the corps and division commanders, and obtaining your approval, I determined to retire during the night of the 30th toward Nashville. The artillery was withdrawn to the north bank during the early part of the night, and at twelve o'clock the army withdrew from its trenches and crossed the river without loss. During the next day, December 1, the whole army was placed in position in front of Nashville.

Information obtained since the above report was written,[1] and principally since the reoccupation of Franklin by our troops, makes the enemy's loss 1,750 buried upon the field, 3,800 disabled and placed in hospitals in Franklin, and 702 prisoners, making 6,252 of the enemy placed *hors de combat*, besides the slightly wounded. The enemy's loss in general officers was very great, being 6 killed, 6 wounded, and 1 captured. It is to be observed that more than half of our loss occurred in General Wagner's division of the Fourth Corps, which did not form part of the main line of defence. This loss arose in two brigades of that division from their remaining in front of the line after their proper duty as outposts had been accomplished, and after they should have taken their positions in reserve, and in the other brigade (Colonel Opdycke's) in its hand to hand encounter with the enemy over the portion of the parapet which had been temporarily lost by the precipitate retreat of the other two brigades. When it became apparent that we should have to fall back from Columbia, orders to rejoin the army were sent to General Cooper, commanding the troops guarding the crossings of Duck River below Columbia at Centerville, both by myself and the Major General commanding, which were obeyed as soon as received, and General Cooper marched for Franklin. Owing to delays in receiving his orders and the time necessary to

[1] The final report was dated December 31, 1864, incorporating and adopting the preliminary report of December 7, preceding.

concentrate his troops, General Cooper could not reach Franklin before its occupation by the enemy, and turned his column direct for Nashville. Arrived at the Brentwood Hills by the Charlotte pike on the night of December 2, and again found the enemy between him and the army. He then marched to Clarksville, where he arrived in safety on the 5th and rejoined my command on the 8th of December. General Cooper deserves great credit for the skill and judgment displayed in conducting his retreat.

APPENDIX B

GENERAL COX'S REPORT

DATED JANUARY 10, 1865

(*Battle of Franklin* [1])

I HAVE the honor to make the following report of the operations of the Twenty-third Army Corps in the battle between the U. S. forces, under Major General Schofield, and the rebel army, under General Hood, at Franklin, Tenn., on the 30th day of November, 1864.

My own division (Third Division, Twenty-third Army Corps) reached Franklin about an hour before daybreak on the morning of the 30th, having marched from Columbia, twenty-two miles, during the night. The division was halted and massed on the ground upon which the battle was fought, and the men were allowed to cook their breakfast whilst the trains which were following filed past into the town. General Schofield, being with the head of the column, after an examination of the means of crossing the army to the north side of the Harpeth River, informed me that the means were so inadequate as to demand his immediate personal attention, and ordered that I assume command of the corps and put it in position to cover the crossing of the remainder of the army to the north bank of the river. The whole command was moving in from Columbia and Spring Hill by the turnpike leading from those places to Franklin, and the enemy was known to be following with his infantry by

[1] O. R., xlv. part i. pp. 349-356.

the same route, his cavalry being chiefly upon the turnpike leading from Lewisburg to Franklin. A reconnoissance of the position as soon as it was light showed that the ground immediately south of the village was almost level, and without any cover from woods or orchards for a distance of nearly a mile from the outskirts of the village, and even for a considerably longer distance on the Columbia pike.

A brick dwelling, belonging to a Mr. Carter, the southernmost one of the town, stands on the west side of the turnpike upon a slight knoll over which the road runs as it leaves the village. This knoll has an elevation of about ten feet above the lower ground around it, and even less above that directly south, the slope there being so slight as to be scarcely perceptible to one approaching from that direction. The crest of this elevation is about two hundred yards in length from right to left, and is divided nearly equally by the Columbia pike. Two other turnpikes diverge from the village going southward, the Lewisburg pike on the left (east) and the Carter's Creek pike on the right (west). A curved line, intersecting these two last-mentioned roads at the edge of the village, crossed each of them upon slight elevations of ground, similar to that at Carter's house on the Columbia pike. This being the only line apparently tenable near the outskirts of the town, and sufficiently short to be occupied in reasonable strength by the two divisions of the corps (the Second being weakened by the absence of the strongest brigade), and it being also substantially the line indicated by the major general commanding upon our approach to the town, I ordered the troops into position upon it, and directed that they throw up breastworks immediately. To completely understand the nature of the field it is, however, necessary to notice that the railroad also passes out of the town toward the southeast, and a little to the left of the Lewisburg pike, and that the Harpeth River, running northwestwardly, is nearly parallel to the railroad and quite near to it for some dis-

Appendix B 313

tance, whilst on our right it opens a considerable space between it and the Carter's Creek pike. Upon the north bank of the Harpeth and near the left of our line, as indicated, is a fort, erected some two years since (Fort Granger), which commands a stretch of the river to the left, and also a cut of the railroad, through which troops might advance under cover toward the left of our line. Reilly's brigade (First), of my own division, was placed with its right resting upon the Columbia pike, its front line consisting of the 100th Ohio and 104th Ohio Volunteers, its second line of the 12th and 16th Kentucky, and the 8th Tennessee Volunteers. Its left extended somewhat beyond a cotton gin, which stood in a slight angle of the line about one hundred yards from the Columbia Turnpike. The Second Brigade (Colonel J. S. Casement, 103d Ohio, commanding) extended the line from Reilly's left to the Lewisburg pike, the 65th Indiana, 65th Illinois, and 124th Indiana Volunteers forming his first line, and the 5th Tennessee Volunteers in the second line. The Third Brigade (Colonel I. N. Stiles, 63d Indiana, temporarily commanding) continued the line from Casement's left to the Harpeth River, the 128th Indiana, 63d Indiana, and 120th Indiana Volunteers in the first line, and 112th Illinois Volunteers in the second line.

Upon the right of the pike I directed Brigadier General Ruger, commanding Second Division, Twenty-third Army Corps, to put his division upon the line indicated, reaching as far to the right as he could firmly hold the line. He accordingly placed Strickland's brigade (Third) upon his left, being immediately on the right of the Columbia pike, the 50th Ohio and 72d Illinois Volunteers in the first line, and the 183d Ohio and 44th Missouri Volunteers in the second line. Moore's brigade (Second) was placed on the right of Strickland's, and in order to cover the Carter's Creek pike was deployed in one line in the following order: 80th Indiana, 23d Michigan, 129th Indiana, and 111th Ohio Volunteers, numbering from right to left, as in the cases of all the other brigades mentioned above. Moore's line being still weak on account of its extent, General Ruger ordered fifty men of the 183d Ohio (Strickland's second line) to report to him, and they were placed by Colonel Moore between the 129th Indiana and 23d Michigan.[1] By noon a tolerably good line of breastworks had been erected along the front described, and in a portion of the line a slight abattis had been constructed. A small locust grove and some fruit trees in front of Ruger's division had been used for this purpose, and some Osage orange hedges about a small enclosure in front of Stiles's brigade on the left had also been made good use of. One line of this hedge parallel to Stiles's left front was slightly thinned out and left standing, and in the end proved most useful. The remainder of the hedge was used along the front of the Third Division, but there was not sufficient material near at hand to make the line continuous, nor was there time to stake it down, so that it amounted simply to a slight obstruction of small branches and twigs that could offer no serious obstacle to an advancing enemy, except as the thorny nature of the Osage orange made it an unexpectedly troublesome thing to handle or remove under fire. The artillery of the corps had been moved to the north side of the river early in the morning, under the direction of Lieutenant Colonel Schofield, Chief of Artillery, and a portion of it placed in the fort.

As the troops of the Fourth Corps came in, later in the forenoon, four batteries from that corps were ordered to report to me, and I assigned them positions as follows: 1st Kentucky Light Artillery, four guns, on the left of the Columbia pike, in the line of the 100th Ohio Infantry; 6th Ohio Light Artillery, two guns at the left of the cotton gin, and two guns on the left of the Lewisburg pike; 20th Ohio Light Artillery, four guns, on the right of the Columbia pike, just west of Carter's house; and Battery B, Pennsylvania Volunteers, at the Carter's Creek

[1] For a corrected statement of the array of Moore's troops, see p. 55, *ante*.

pike.[1] Although not strictly in the order of occurrence, it will tend to greater clearness to add that about three o'clock in the afternoon, when an attack by the enemy in force had become more immediate, other batteries of the Fourth Corps were placed in position by Lieutenant Colonel Schofield and Captain Bridges, Chiefs of Artillery of the Twenty-third and Fourth Corps, respectively, viz.: Battery M, 4th U. S. Artillery, and Battery G, 1st Ohio Light Artillery, were thus placed near the left of Stiles's brigade, Third Division, Twenty-third Army Corps; Battery A, 1st Ohio Light Artillery, was placed in reserve near the Columbia pike; and Bridges's Battery, Illinois Light Artillery, was placed near the centre of Strickland's brigade, Second Division, Twenty-third Corps. About noon, some appearance of the enemy's cavalry being reported on the Carter's Creek pike, I called the attention of the commanding general to the fact that Ruger's division could not reach any secure point at which to rest on the right, and shortly after Brigadier General Kimball, commanding First Division, Fourth Corps, reported to me by order, and I directed him to go into position on General Ruger's right, filling the space between the Carter's Creek pike and the river.

At two o'clock I received orders to withdraw the command to the north bank of the river at six o'clock, in case there should be no attack by the enemy. At this time nearly the whole of the trains and Wood's (Third) division of the Fourth Corps, had crossed the Harpeth. Shortly after, Brigadier General Wagner, commanding Second Division, Fourth Corps, presented in person his orders to report to me and act under my orders. He informed me that one brigade (Opdycke's) of his division was already within the lines, and that the other two, with a section of artillery, had been acting as rear guard for the army and were then some two miles at the

[1] Two lines in the description of the positions of artillery were, by a clerical error, omitted in the report forwarded to Washington, and are inserted from my retained copy of the original.

rear, where the Columbia pike passes through a high range of hills before reaching the plateau on which the village stands; that his orders then were to hold the enemy back until they developed a heavy force manifestly superior to his own, and then slowly retire within my lines. I directed Opdycke's brigade to be placed on the right of the Columbia pike, about two hundred yards in rear of our centre, as a general reserve; that the orders under which General Wagner was then acting as to the two brigades serving as rear guard should be carried out, and that when the troops were withdrawn within the lines they should be placed in position near Opdycke's brigade and held in reserve awaiting further orders and in readiness to support any part of the line. At three o'clock the two brigades of Wagner's division in front had fallen back to a position about half a mile in front of the lines, and reported the enemy developing in force in their front, whilst they opened upon the rebels with the section of artillery which was with them. The order was then reiterated to General Wagner to withdraw the brigades whenever the enemy appeared to be advancing in decidedly superior force, without allowing his troops to become seriously engaged. General Wagner was at that time in person upon the Columbia pike near the Carter house, where my headquarters had been during the day.

A slight depression beyond the lines held by Wagner's advanced brigades prevented the enemy from being seen from our lines till about four o'clock, when the officers on the skirmish line reported him advancing in several lines and in very great force. Almost simultaneously with this report the two brigades of Wagner's division in front opened a brisk musketry fire, and part of them were seen making a barricade of rails, etc., apparently with a view of endeavoring to make a stand there, though the section of artillery retired leisurely within our lines. Before an order could reach them they were so hotly engaged that they could not be withdrawn in order. The enemy wasted no time in firing, but charged them, and, being enveloped on the flanks, the two brigades, after a short and brave though useless struggle, broke to the rear in confusion. The momentary check at the centre brought the right wing of the enemy farther forward, and they came on at a double-quick with trailed arms, some pieces of artillery advancing and firing between brigade intervals. As soon as they were seen the batteries on our left opened upon them, as well as the guns in Fort Granger, and as they advanced into rifle range of our infantry, Stiles's and Casement's brigades opened fire also. The rebel lines could now be plainly seen, as well as the general disposition and apparent purpose of their movement. Their heaviest masses were advancing on the line of the Columbia pike, reaching quite to the river on our left, the two points of assault at that time being apparently our centre and our extreme left; the latter being the point nearest to our bridges, which were necessarily much nearer that flank (one of them being the railroad bridge), and that being the line of movement by which they would most rapidly have cut us off from crossing the river had our lines been broken. The extreme left was the portion of our main line first warmly engaged. The enemy endeavored to pass up the railroad cut above mentioned, but were enfiladed not only by the guns in Fort Granger but by Battery M, Fourth U. S. Artillery, and driven from that shelter. Their lines on either side, however, advanced steadily. On reaching the Osage orange hedge in front of Stiles's left, they first endeavored to force their way through it and pull it aside. The tough and thorny nature of the shrub foiled them in this, and they attempted to file around the hedge by the flank, and under a terrible, withering fire from Stiles's and Casement's brigades and the batteries on that flank. They soon abandoned this effort, and most of those remaining unhurt lay down behind the hedge, and after keeping up a desultory fire for a time, straggled to the rear, singly and in small squads.

In front of Stiles's right and Casement's left, the obstructions being fewer and more insignificant, the enemy advanced rapidly and in good order, though suffering very severely, up to the breastworks, and made desperate efforts to carry them. Their officers showed the most heroic example and self-sacrifice, riding up to our lines in advance of the men, cheering them on. One general officer (Adams) was shot down upon the parapet itself, his horse falling across the breastwork. In all this part of the line our men stood steadily without flinching, and repulsed the enemy, inflicting terrible loss upon him and suffering but little in return. Meanwhile, in the centre, the enemy gained some temporary advantage. When the two brigades of Wagner's division, Fourth Corps, broke, the enemy were close upon them and followed them in, overtaking and capturing considerable numbers of the fugitives. Our own men in the lines along the centre were restrained from firing, in order not to injure those who were retreating, and the enemy were thus enabled to come up to the breastworks pell-mell with Wagner's men, without suffering loss or being seriously exposed to fire. Immediately upon the pike the crowd of the retreating troops and the advancing enemy overwhelmed the men at the breastworks there, and a portion of the right of Reilly's brigade (Third Division) and most of Strickland's brigade (Second Division) broke from the first line. This was not due altogether to the pressure upon their immediate front, but partly also to the fact that the orders given by their officers to the rear of Wagner's division coming in from the front to rally at the rear were supposed by many of the men in the lines to apply to them also. When the two brigades of Wagner's were first seen to be compromised by getting seriously engaged, as a provision against danger in the centre I had ordered Opdycke's brigade to be ready to charge up to the lines instantly, if there should be any confusion there. This brigade was now ordered up, and came up the turnpike in the most gallant manner; Reilly's rushed forward at the same moment.

Major General Stanley, commanding Fourth Corps, who had been ill during the prior part of the day, came on the field on hearing the sound of battle, and arrived in time to take an active part in the effort to rally Wagner's men, but was soon wounded and his horse shot under him. The most strenuous efforts were made by all officers along that part of the line to rally the men, and were so far successful that the line was quickly restored on the left of the turnpike, and after a sharp struggle on the right of Strickland's brigade also, though the enemy continued to occupy in some force a portion of the outside of the parapet on Strickland's left for a distance of about one regimental front. Opdycke's brigade occupied the second line, which at that point was not over twenty-five yards in rear of the first, and under cover of the smoke strengthened a barricade and breastwork which had been before there. The 175th Ohio Volunteers, a new regiment, unassigned to a brigade, had reported early in the afternoon by direction of the commanding general, and was by me temporarily assigned to Reilly's brigade and placed in reserve. It also advanced with the rest of the supporting troops and did good service, behaving with great steadiness and courage. The attack extended toward our right to the Carter's Creek pike. The enemy, being apparently satisfied of the impracticability of advancing again upon our left for the reason before stated, pressed farther to our right, and especially after they had seemed to have gained some advantage in the centre, their efforts there and upon their own left were redoubled. Colonel Moore's brigade held its ground firmly, and although it was in imminent danger at the moment when the centre wavered, repulsed a determined assault, and preserved its line intact throughout the battle. The condition of the atmosphere was such that the smoke settled upon the field without drifting off, and after the first half hour's fighting it became almost impossible to discern any object along the line at a few yards' distance. This state of things appeared to have deceived Colonel Strickland in regard to his line, as he reported the first line completely reoccupied along his entire front after the repulse of the enemy's first assault, whilst in fact a portion of it at his left was not filled by our troops, and Colonel Opdycke, not being personally acquainted with the lines, was not aware for some time that he had not reached the first line in Colonel Strickland's front, where the outbuildings of Carter's house prevented the line from being distinctly seen from the turnpike even if the smoke had not formed so dark a covering.

After a short lull the attack was resumed by the enemy with the same audacity and determination as before, and, Strickland's brigade suffering considerably, and being reported by him a good deal weakened, I withdrew the 112th Illinois Volunteers from the second line of Stiles's brigade on the extreme left, and ordered it to report to Colonel Strickland and to aid in re-establishing the line in his front. It was led forward very gallantly by Lieut. Colonel Bond commanding, who was wounded in the advance. The smoke and growing darkness deceived also the enemy, who apparently supposed they had gained full possession of our lines in the centre, and continued to push in fresh masses of troops, only to be destroyed or captured, for very few went back, insomuch that prisoners captured continually expressed the utmost surprise, declaring that they supposed and had been informed that our lines were occupied by their troops which had assaulted before, and of whom nothing since had been seen. The ditches in front of the whole line of the corps, and particularly in the centre, contained many of the enemy who were unable to get back, and who at the first opportunity surrendered and came over the breastworks as prisoners. The assaults on the centre, extending considerably to the right of the Columbia pike and involving Moore's brigade more or less, were obstinately repeated until nightfall, and even as late as nine o'clock attacks were made, which were, however, easily repulsed, and the enemy withdrew the remnants of his shattered lines to the position occupied at the opening of the battle by Wagner's division, in advance of our lines about eight hundred yards. Alarms occurred frequently until eleven o'clock, and frequently caused a general musketry fire on both sides from our centre toward the right; but I found no evidence that any real attack was made at so late an hour, the demonstrations being manifestly made by the rebels to discover whether our lines were being abandoned during the evening.

At midnight, all being quiet in the front, in accordance with orders from the commanding general, I withdrew my command to the north bank of the river, leaving a skirmish line in the earthworks an hour later, when they also were withdrawn. The whole movement was made without interruption or molestation from the enemy, the Third Division moving by the left flank and crossing the river upon the railroad bridge, which had been planked, and the Second Division (with Opdycke's brigade of the Fourth Corps) moving through the town and crossing by a wagon bridge a little below the railroad crossing. Upon reaching the north bank I took up the line of march with my own division for Brentwood in advance of the army, by command of General Schofield. General Wagner rallied the two brigades of his division at the river, but they were not again brought into action. Kimball's division of the Fourth Corps was to some extent engaged upon its extreme left in the late attacks, which reached to and somewhat beyond the Carter's Creek pike, and that command also suffered somewhat from the diagonal fire of the enemy upon Ruger's division of this corps. This, however, I state from my own casual observation alone, as I took no control of the troops of the Fourth Corps (except Opdycke's brigade) after General Stanley came upon the field, and have no official report of their part in the engagement. The casualties of the corps during the engagement are reported to me as follows.[1]

These lists were made up soon after the engagement, and I am convinced that corrected ones, when procured, will show a considerable diminution in the list of the missing. The loss of the enemy we are enabled to approximate with some accuracy from the public admissions from their commander as well as from the statements of prisoners, our own examination of the field after it again came into our possession, and the statements of citizens and hospital attendants remaining in Franklin. From all these sources the testimony is abundant that the rebels lost 6 general officers killed, 6 wounded, and 1 captured; that they buried 1,800 men on the field, and that 3,800 were wounded. The number of prisoners captured by this corps was 702. Thus, without estimating the prisoners taken by any part of the Fourth Corps, or the stragglers and deserters, who are known to have been numerous, the enemy's loss was not less than 6,300. The attack was made by Stewart's and Cheatham's corps of Hood's army, Lee's corps being in reserve, and it is only repeating what is proven by the concurrent testimony of all officers and men of the rebel army who were captured, when I assert that the two assaulting corps were so weakened in numbers and broken in morale in this engagement as to lose for the rest of the campaign the formidable character as opponents which these veteran soldiers had before maintained. Their remarkable loss in general officers attests sufficiently the desperate efforts to break our lines and the heroic bravery of our own troops, who repulsed their repeated assaults.

. . . .

The transmission of this report has been delayed by reason of waiting for reports of subordinate commanders, and the whole are now submitted.

[1] See table, chap. xvi., *ante.*

APPENDIX C

GENERAL WOOD'S REPORT

DATED JANUARY 10, 1865

(Part relating to Battle of Franklin[1])

HEADQUARTERS FOURTH ARMY CORPS,
HUNTSVILLE, ALA., *January* 10, 1865.

IN pursuance with orders received from the headquarters Department of the Cumberland to report the operations of the corps from the time it was detached from the main army of the Military Division of the Mississippi, in the latter part of October, to its arrival at Nashville, on the 1st of December ultimo, I have the honor to submit the following. . . .

On arriving at Franklin the Twenty-third Corps had taken position in the suburbs of the village, with its left resting on the river above the town and its right extending across and west of the turnpike road. The First Division of the Fourth Corps (Kimball's) was posted on the right of the Twenty-third Corps, with its right flank resting on the river below the town. Intrenchments were at once thrown up by the Twenty-third Corps and Kimball's division, of the Fourth Corps. The Third Division of the Fourth Corps arrived next, and was ordered to cross the river and take post on the north side. This was done. Wagner's (the Second Division), which was marching in rear, was ordered to halt on a range of hills nearly two miles south of the town and deploy his command to hold the enemy in check should he attempt to press us. In the mean time the transpor-

[1] O. R., xlv. part i. pp. 119-126.

tation was being passed rapidly across the river. At 12 M. General Wagner reported the appearance of the enemy in heavy force in his front, and later he reported that the enemy was evidently making preparations to attack him in force. The position General Wagner then held was entirely too extensive to be covered by one division, and as the country was open on both flanks and favorable to the movement of troops, the position could be readily flanked; hence General Wagner very judiciously determined to retire his command nearer to the town. He posted two brigades, Conrad's and Lane's, across the pike, with their flanks slightly refused, about a third of a mile south of the intrenched position of the Twenty-third Corps. The other brigade, Opdycke's, was sent inside of our main works, — a most fortunate disposition of this brigade, as the sequel of this narrative will show. Conrad's and Lane's brigades hastily threw up rude barricades to protect themselves from the coming storm. Their orders were to maintain their position as long as it could be done without becoming too severely engaged, and then retire on the main line. At 4 P. M. the enemy made a vigorous attack on the front of these two advanced brigades, threatening at the same time their flanks with strong columns. Unwilling to abandon their position so long as there was any probability of maintaining it, unfortunately the gallant commanders remained in front too long, and as a consequence, when they did retire, they were followed so closely by the enemy as to enter the works through the break which had been caused by the burst over them of the retiring brigades. The enemy had come on with a terrific dash, had entered our intrenchment, and victory seemed almost within his grasp. Our line had been broken in the centre, two four-gun batteries had fallen into the hands of the enemy, and it seemed that it was only necessary for him to press the advantage he had gained to complete his success. But at this critical moment the gallant, prompt, and ready Opdycke was at hand, calling to his men in a stentorian tone, "Forward to the lines!" and, adding example to command, he, with his bold brigade, with lowered bayonets, rushed forward, bore the exultant enemy back over our intrenchments, recovered the lost guns, and captured nigh 400 prisoners. But this reverse did not seem to discourage the enemy; it seemed rather to add to his determination and increase the vigor of his assaults. On, on, he came, till he made four distinct assaults, each time to be hurled back with heavy loss. So vigorous and fierce were these assaults that the enemy reached the exterior slope of the rude intrenchments, and hand to hand encounters occurred between the courageous combatants across the works; and between the assaults the work of death was not stopped. The undulations of the ground are such as to afford good protection to an attacking force. Under this cover the enemy pressed sharpshooters as near our lines as possible, and kept up a most galling fire. While these vigorous attacks were being made on our centre and left, the right, held by Kimball's division, was also fiercely attacked three times, all of which assaults were handsomely repulsed, with comparatively slight loss to us, but with terrific slaughter of the enemy. At no time did the enemy gain any advantage on this part of our lines. As night approached the enemy desisted from his fierce assaults, and his offensive efforts degenerated into a sharp skirmish fire.

Thus terminated one of the fiercest, best contested, most vigorously sustained passages at arms which have occurred in this war.

. . . .

When the enemy had temporarily broken our centre, Major General Schofield, commanding the forces in the vicinity of Franklin, under the apprehension that our forces engaged on the south side of the river might be compelled to pass to the north side, ordered the following disposition, which was made, of the Third Division (then in reserve) of the Fourth Corps, with a view to covering the withdrawal of our troops, should it become necessary: Beatty's brigade was deployed on the north bank of the river above the town; Streight's brigade along the bank of the river immediately opposite the town; and Post's brigade on the bank of the river below the town. Fortunately the exigency for which this disposition was made did not occur in the progress of the contest, but the brigades retained their positions to cover the withdrawal at night, which had been ordered before the occurrence of the attack. To prepare for the withdrawal and retirement toward Nashville, the trains were started before nightfall of the 30th. At midnight the troops on the south side of the river began to withdraw from the lines and pass to the north side of the stream; this work was rapidly and successfully accomplished. The enemy probably suspected what was going on, but did not attempt to interfere with the movement. The Third Division of the Fourth Corps had been designated to move in rear and cover the retrograde movement. By 3 A. M. of the 1st of December all the troops had been withdrawn from the south side of the river, and the bridges were fired. So soon as the conflagration was so far advanced as to insure its being complete, the Third Division commenced to withdraw, and by 4 A. M. the whole of it was on the road. As the flames rose from the bridge and communicated fully to the enemy our movements, he opened a heavy cannonade, which fortunately did no injury, and was soon suspended; otherwise the enemy did not attempt to molest us. The rear of the command reached Brentwood, nine miles north of Franklin, at 9 A. M.

. . . .

In conclusion it is proper that I should remark that during the operations briefly sketched in this report the corps was commanded by Major General Stanley. He was wounded in the battle at Franklin, and was compelled by the casualty to relinquish command of the corps before he could submit a report of its operations; hence the duty has devolved on me.

APPENDIX D

GENERAL STANLEY'S REPORT

DATED FEBRUARY 25, 1865

(Part relating to Battle of Franklin [1])

GENERAL KIMBALL's division reached Franklin soon after nine o'clock and took up position on the right of the Twenty-third Corps, the right flank of the division resting on the Harpeth below Franklin. The line selected by General Schofield was about a mile and a half in length, and enclosed Franklin, resting the flanks upon the river above and below the town. The trains were all crossed over to the north side of the Harpeth; Wood's division was also crossed and posted to watch the fords below the place. Colonel Opdycke reached the heights two miles south of Franklin at 12 M. He was directed to halt on the hills to observe the enemy. Croxton's brigade of cavalry was steadily pushed back by the enemy's infantry column on the Lewisburg pike, and at one o'clock General Wagner reported heavy columns of infantry approaching on the Columbia and Lewisburg pikes. General Wagner was instructed to fall back before the advance of the enemy, observing them. About the same time word was received that the rebels were trying to force a crossing at Hughes's Ford, two miles above Franklin.

From one o'clock until four in the evening the enemy's entire force was in sight and forming for attack, yet in

[1] O. R., xlv. part i. pp. 112-118.

view of the strong position we held, and reasoning from the former course of the rebels during this campaign, nothing appeared so improbable as that they would assault. I felt so confident in this belief that I did not leave General Schofield's headquarters until the firing commenced. About four o'clock the enemy advanced with his whole force, at least two corps, making a bold and persistent assault, which, upon a part of the line, lasted about forty minutes. When Wagner's division fell back from the heights south of Franklin, Opdycke's brigade was placed in reserve in rear of our main line, on the Columbia pike. Lane's and Conrad's brigades were deployed — the former on the right, the other the left of the pike — about three hundred yards in front of the main line. Here the men, as our men always do, threw up a barricade of rails. By whose mistake I cannot tell, it certainly was never a part of my instructions, but these brigades had orders from General Wagner not to retire to the main line until forced to do so by the fighting of the enemy. The consequence was that the brigades stood their ground until the charging rebels were almost crossing bayonets with them, but the line then broke — Conrad's brigade first, then Lane's — and men and officers made the quickest time they could to our main lines. The old soldiers all escaped, but the conscripts being afraid to run under fire, many of them were captured. Conrad's brigade entered the main line near the Columbia pike, Colonel Lane's several hundred yards to the right of the pike. A large proportion of Lane's men came back with loaded muskets, and, turning at the breastworks, they fired a volley into the pressing rebels now not ten steps from them. The part of the Twenty-third Corps stationed in the works for a distance of about three [hundred] or four [hundred] yards to the right of the Columbia pike, and which space took in the 1st Kentucky and 6th Ohio Batteries, broke and ran to the rear with the fugitives from Conrad's brigade. To add to the disorder the caissons of the two batteries galloped rapidly to the rear, and the enemy appeared on the breastworks and in possession of the two batteries, which they commenced to turn upon us. It was at this moment I arrived at the scene of disorder, coming from the town on the Columbia pike; the moment was critical beyond any I have known in any battle — could the enemy hold that part of the line, he was nearer to our two bridges than the extremities of our line. Colonel Opdycke's brigade was lying down about one hundred yards in rear of the works. I rode quickly to the left regiment and called to them to charge; at the same time I saw Colonel Opdycke near the centre of his line urging his men forward. I gave the colonel no order, as I saw him engaged in doing the very thing to save us, viz., to get possession of our line again. The retreating men of Colonel Conrad's brigade, and, I believe, the men of the Twenty-third Corps, seeing the line of Opdycke's brigade start for the works, commenced to rally. I heard the old soldiers call out, "Come on, men, we can go wherever the general can," and, making a rush, our men immediately retook all the line, excepting a small portion just in front of the brick house on the pike. A force of the rebels held out at this point, and for fifteen or twenty minutes, supported by a rebel line fifty yards to the rear, poured in a severe fire upon our men. So deadly was this fire that it was only by the most strenuous exertions of the officers that our men could be kept to the line; our exertions, however, succeeded, and in twenty minutes our front was comparatively clear of rebels, who fell back to the position formerly held by the two brigades of the Second Division in the commencement of the fight, from whence they kept up a fire until midnight, when we withdrew. Just after the retaking of the line by our troops, as I was passing toward the left to General Cox's position, my horse was killed; and no sooner had I regained my feet than I received a musket-ball through the back of my neck. My wound, however, did not prevent my keeping the field, and General Cox kindly furnished me a remount. The rapidity of the firing made it very difficult to keep up the ammunition, the train being some two miles distant on the road to Nashville when the battle commenced, and our greatest danger at one period of the battle was that we would exhaust our ammunition. One hundred wagon-loads of ammunition, artillery and musket cartridges, were expended in this short battle, belonging to the ordnance train of the Fourth Army Corps; this train, however, supplied in great part the wants of the Twenty-third Corps.

After the first great attack and repulse the enemy made several feeble demonstrations, and until nine o'clock in the evening formed and advanced upon the Columbia pike three or four times. I think these movements were made to keep us from moving, or to ascertain the very moment we left. At the commencement of the engagement word came that the enemy's cavalry had forced a crossing at Hughes's Ford, and calling upon me for support for our cavalry. General Wood was directed to send a brigade, and General Beatty's brigade had started, when information came that our cavalry had driven the rebels back and the reinforcements would not be needed. General Kimball's division, holding the extreme right of the line, had comparatively an easy thing of this fight; being well posted behind breastworks, their volleys soon cleared their front of rebels. One brigade, Colonel Kirby's, only had the opportunity to fire one volley, and this was a very effective one, at a rebel brigade which endeavored to move obliquely across our front to gain the right bank of the river. It having been determined to withdraw the troops to Nashville, they were directed to leave the line at midnight, the flanks withdrawing first and simultaneously, the pickets to be withdrawn when all the troops had crossed. Some villain came very near frustrating this plan by firing a house in Franklin; the flames soon spread, and the prospect was that a large fire would occur, which, lighting up objects, would make it impossible to move the troops without being seen. My staff officers and General Wood's found an old fire engine, and getting it at work, the flames were soon subdued, and the darkness was found to be increased by the smoke. At midnight the withdrawal was made successfully, although the enemy discovered it and followed our pickets up closely.

General Wood's division remained on the north side of the Harpeth until four o'clock in the morning as rear guard, destroying the bridges before he left. The enemy indulged in a furious shelling as soon as they found we had left. In the fight of the day before their artillery had not come up, and but two batteries were used upon us. These two batteries threw shells into the town during the entire fight.

. . . .

Although Brigadier General J. D. Cox was not in my command, he was my close neighbor in the battle of Franklin, and I take this opportunity to express to him my thanks for his gallant help at that time.

APPENDIX E

STATEMENT OF COLONEL DOW

NOVEMBER 29, 1864. Was left as Inspector, Third Division, Twenty-third Corps, in charge of pickets at Duck River, when the division was moved to Spring Hill and Franklin. Had with me the pickets actually on duty and 12th and 16th Kentucky as supports, with two ambulances and one wagon loaded with ammunition. The skirmish line was a long one, and consisted of some hundreds of men besides the supports.

Remained till midnight, as ordered, then relieved the skirmishers, and drew them in, and started with them for Spring Hill and Franklin to overtake the division. Passed Spring Hill just as the sun was rising, and overtook the rear of the Fourth Corps a mile or so north of Spring Hill.

Met General T. J. Wood, who was at what I supposed was the rear of the column. After informing him who I was and the command with me, General Wood made a remark which I remember. He said, "Well, Major, do you think the Lord will be with us to-day?" I marched in rear of all the column of which I knew anything, some four or five miles, when the enemy's cavalry made an attack on our right, and had evidently been trying to reach the train. They were occupying higher ground to the east of the pike, perhaps a thousand yards away. General Wood told me to take my command and go over the fence, and drive those fellows off. I obeyed him, and did so, and we drove them off easily. A section of artillery took part in this affair from a position, say a quarter of a mile ahead of us, on the pike.

We passed some wagons which had been destroyed by the dash of the enemy's cavalry, but whether then or at another time that morning, can't be sure. They (the wagons) were already nearly burned up when we saw them.

We reached Franklin about noon, when I reported my command, which was ordered into second line of Reilly's brigade to which it belonged. I distinctly remember that the 16th Kentucky lay with its right at the pike in Reilly's second line, and the 12th lay on its left. After these joined their command, I returned to staff duty. Being greatly fatigued, I threw myself down somewhere after I got my dinner, and slept till about the time of the enemy's advance to open the battle.

Being told they were forming, I went to the lines on the pike, where the pike passed through the breastwork. I saw the enemy advancing. I saw the battery that had been with Wagner's two brigades trot leisurely through the works. Don't remember anything about Wagner himself. Saw the break at the front, and saw Wagner's men coming in in confusion. Several of the staff were together, and when we called for our horses, the orderlies had disappeared with them. Some went to bring them up again or to join the General (Cox), who was at that time farther to the left.

The first incident I distinctly remember after this was seeing the 16th Kentucky go forward in fine style along the left of the pike to the works from which the 100th Ohio had been driven with the crowd of fugitives that came in. They regained the line. Opdycke's brigade came forward, and Generals Stanley and Cox were on the pike with them. Captain Tracy was with General Cox, and I joined in the effort to rally the men who had come in in confusion or who had left the line. General Stanley was wounded in the first mêlée, and his horse shot also. I saw his wound in the neck, and saw him remounted on one of General Cox's horses. General Cox's headquarters were close to the pike on its left, in rear of

Reilly's right, during the remainder of the battle, and from there he and the staff went to do any duty that called.

I don't remember seeing Wagner during the battle. Stanley exercised no command while there, but rallied the men. I was along the line on duty, more or less, during the engagement. I carried the order to the 112th Illinois to go to the right and attempt to get Strickland's line up to the original breastworks. I led the regiment across, halted them near the pike, and explained the thing expected of them; then took them to Strickland's line, and we got over his barricade and crawled forward to the front line, a distance of three or four rods. It was already dark or dusk, and I can't be entirely certain as to the distance back of the Carter house where we got over the works. It was a short distance west of the pike. We crept forward to the front, and as we rose were met with so heavy a fire that we were obliged, after a little, to go back to Strickland's new (or second) barricade.

I knew at the time that part of the 104th Ohio got over the works in their front soon after our effort in Strickland's line, captured the enemy lying in the trenches, and cleared the front there if not in front of Strickland. For reasons I will state presently, I am confident none of the enemy's force stayed at the outside of our breastworks long after dark, and that they retired under cover of darkness, all who were able to, to positions near where Wagner's two brigades had been.

Some time in the evening I was sent by General Cox to General Schofield's headquarters on the north side of the river. I found him and General Stanley together. They were at a house, say half or three quarters of a mile from the river, where I understood General Schofield fixed his headquarters about dark. I had been sent to arrange with Colonel Hartsuff, Inspector of the Corps, for the skirmish line which was ordered to be left when the line should be withdrawn. When the request had

Appendix E 335

first been sent to General Cox to have me go to meet Hartsuff, General Cox did not permit me to go, as we were busy at the front, but later in the evening I went over at the General's direction.

I am confident a good many of the 100th Ohio went back to the line after the second line advanced. The works east of the pike were full of men, three or four deep, showing that the ranks were practically doubled, or more. Indeed, there was quite a crowd all along there.

When the command moved out of the works, near midnight, I was left in charge of the skirmish line, which was detailed to occupy them for another hour. I went over the works, and walked some distance out in front. No enemies were there but those disabled or dead, and the cries of the wounded for help were very distressing. At that time I heard no signs of any force on either side of the pike. There had been occasional sharp volleys all the evening, but these had ceased, and there was no shot fired while our troops were withdrawn, nor during the time I stayed, or while I withdrew the skirmishers. We moved off entirely undisturbed, and overtook the command after daylight.

The foregoing is written at my dictation, and is a correct statement of the facts which I remember as to the battle of Franklin and the events of the preceding night.

(Signed)   T. T. Dow.

APPENDIX F

STATEMENTS OF COLONEL COX

I

CINCINNATI, OHIO, *June* 16, 1881.

MY DEAR GENERAL, — Perhaps my personal recollections of some of the incidents connected with the battle of Franklin may be of service some day in connection with more important testimony, and I jot them down for your use.

After we had established our lines about the town to cover the crossing of our army trains over the Harpeth River, most of the officers of your staff were trying to get a little rest after our hard night march from Columbia and Spring Hill, and were lolling around on the porch and grass in front of the Carter house near to the Franklin pike. Two brigades of General Wagner's division of the Fourth Corps were in line some distance in advance of the main line, they having been the rear guard in the retreat from Columbia. I recollect your having a conversation with General Wagner about the position of his troops and the necessity of their being retired within the main line should Hood attack us, and before there should be any danger of their being engaged where they were. You then rode off toward the left with Captain Tracy, and, I think, Lieutenant Coughlan.

A short time after this Captain Bradley and I were sitting on the steps of the Carter house, when the first premonition to us of battle came in the shape of a shell tearing off a portion of the cornice of the porch and exploding in the yard near by.[1] We immediately ran to our horses, preparing to take the direction you had gone, down the line, and as we crossed the road saw a staff officer come dashing in from the front, up to where General Wagner was, near the opening in the main line on the pike, and we very naturally stopped to hear what he had to say. He reported that the enemy was moving forward in line of battle, evidently in force, and would overlap the two brigades in front on both flanks, and he did not think they ought to remain there any longer. "Go back," said General Wagner, "and tell them to fight, — fight like hell!" Bradley and I at once remonstrated with General Wagner, saying that such an order we were sure was not in accordance with the understanding had with you. But he simply reiterated it more forcibly, if possible, than before, and the staff officer rode off to the front. Bradley and I then started off to report this strange affair to you. Before it could be corrected, however, the mischief was done, and Wagner's two brigades came in only as they came with the enemy on their heels.

After the break in our line on the pike (caused by Wagner's men and the enemy mounting our works at the same time and so obliging that part of our corps to withhold its fire) was re-established by General Opdycke's brigade, General Stanley rode up to the front at this point; but he had scarcely done so (and before he had time to view the field or give or suggest an order) when he was wounded and his horse shot. Captain Tracy then dismounted from your dun horse which he had been riding, and gave it to Stanley, who immediately mounted and rode off through the village. This was the only time I saw General Stanley that day until I went over the river in the evening to see General Schofield at his head-

[1] A veranda on the south side of the ell of the Carter house faced toward our front. A cannon shot passed under this veranda, and through the ell. The holes of entrance and exit are still to be seen.

quarters. Of the great battle itself, it is hardly proper for me to say anything here.

When the many desperate charges in column of the enemy had been repulsed and the firing along the line had so far subsided as to indicate that Hood had given up the fight for the night, I went by your direction over the river to General Schofield's headquarters to report the position at the front, and suggest to General Schofield that you were so well satisfied with the punishment Hood had received, that you urged the holding of the lines and the reinforcement of them from Nashville that night, so that we might take the offensive in the morning, and that you were willing to guarantee our success from that point, with your head.

At General Schofield's headquarters, besides his usual staff, I saw General Stanley and several officers. After making my report and delivering your message to General Schofield, he said to me in the presence of all there, "Go back and tell General Cox that he has won a glorious victory, and however much his suggestions weigh with me, my orders from General Thomas are to fall back to Nashville as speedily as possible, and it must be done. Therefore, so soon as the enemy withdraws sufficiently and General Cox thinks it safe, tell him to put the whole command in motion and cross the river."

We crossed the river that night late, having been kept in position an hour or so by the burning of a barn in town that so illuminated our movements that we were obliged to wait and see it out.

Yours,
(Signed) THEO. COX.

GENERAL J. D. COX.

II

CINCINNATI, *October* 2, 1889.

MY DEAR GENERAL,—I remember very distinctly that our headquarters at Franklin, November 30, 1864, were fixed at the Carter house when we first reached the town early in the morning. After it was determined that the Twenty-third Corps must remain there and defend the position until bridges could be made over the Harpeth and all our army trains crossed to the north bank, a few of our most necessary tents were taken out of the wagons and put up on the slope of the hill just north of the house. During the day and up to midnight, members of your staff and orderlies were always in that vicinity. The tents, however, were removed about the time the battle opened, and I did not see them again till we got settled down at Nashville.

A short time after the line had been restored, Captain Hentig, A. Q. M., reported that stragglers had taken possession of the old county bridge to such an extent as to block his trains, and you sent me down there to help right matters and put a guard upon the bridge. While in that vicinity I saw General Wagner making efforts to rally what I suppose was a part of his broken and demoralized command. He was very excited and demonstrative. In his pleading and remonstrating with his men, I have the most vivid recollection of his using these words: "Stand by me, boys, and I'll stand by you!"

Upon my stating to General Wagner my orders, he volunteered to place a guard on the bridge, and see that no more trouble occurred between the men and the crossing trains.

Ever truly,
(Signed) THEO. COX.

GENERAL J. D. COX, Cincinnati.

INDEX

N. B. — The rank of officers engaged at Franklin is that which they bore at the time.

ADAMS, JOHN, Brig. Gen. com'g Brigade, 88. His charge and death, 127, 128.
AMMUNITION. Twenty-third Corps trains sent over the Harpeth, 50. Supply from Fourth Corps trains, 50, note.
ARTICLES OF WAR, Interpretation of, 274 *et seq.*
ARTILLERY positions at Franklin, 55–59, 75, 84, 108, 175, note. Withdrawal, 186 *et seq.* First orders, 267.
ASHBURY, Capt. com'g Schofield's cavalry escort, 34, note.
ATWATER, FREDERICK A., Capt. com'g 51st Illinois, 79, note.
ATWATER, MERRITT B., Major com'g 42d Illinois, 78, note, 252.
AYERS, STEPHEN C., Surgeon. Hospital work, Incident, 185, note.

BALDWIN, A. P., Lieut. com'g Battery, 110. At the Cotton-Gin, *ib.* Withdrawal, 187.
BALDWIN, BYRON C., Color Serg't. Leads a rally, 109.
BARNES, MILTON, Lieut. Col. com'g 97th Ohio. At outpost, 78, 79, note. Re-organization, 250.
BARR, ANDREW J., Lieut. Col. 44th Missouri, 57. Position at Franklin, *ib.* Report, 118, note. Breastworks, *ib.*, 240. Regiment transferred, 210, note.
BATE, WILLIAM B., Maj. Gen. com'g Division, 88. Formation for attack, 132. Convergent advance, 133. Casualties, 135. Line laps Brown, 152.
BATES, EDWARD P., Capt. com'g 125th Ohio. In Opdycke's charge at centre, 116. New barricades, *ib.*, 238.
BEATTY, SAMUEL, Brig. Gen. com'g Brigade, 168.
BEAUREGARD, G. T., General. His command, 4. Relations to Hood, 5, 6. Judgment on battle of Franklin, 12, 197.
BENNETT AND HAIGH, History 36th Illinois, cited, 241.
BIEGHLE, ALEXANDER W., Lieut. com'g ambulances, 182.
BOND, EMERY S., Lieut. Col. com'g 112th Illinois. Reinforces Strickland, 161. Effort to retake main line, *ib.* Wounded, 162. Later career, 162, note. Second line, 240.
BOSTICK PLACE. Topography, 47. Battery there, 132. Bate's line, *ib.*
BRADLEY, D. C., Lieut. & A. D. C. Wagner's orders, 106.
BRADLEY, LUTHER P., Brig. Gen. com'g Brigade, wounded at Spring Hill, 64. Wagner's characteristics, 231.
BRADSHAW, ROBERT C., Colonel 44th Missouri. Wounded, 57, 117, note, 240. Regiment transferred, 210, note.
BRANTLY, WILLIAM F., Brig. Gen. com'g Brigade, 88. Attack, 153.
BRIDGES at Franklin, 45. Repair of County Bridge, 50. Railway

bridge, 51. Retirement of troops, 188 et seq.

BRIDGES, LYMAN, Captain and Chief of Artillery, 57. Batteries reported to Cox, ib., 58, 82, 260, note. Position of batteries, 108, 117, note, 132. Withdrawal, 187. Understanding of orders, 268.

BROWN, JOHN C., Maj. Gen. com'g Division, 88. His advance, 93, 104. Attacks Ruger, 132. Lines of attack, 152, 155. Wounded, 165.

BROWN, JOHN H., Capt. 12th Kentucky. At the Cotton-Gin, 111.

BROWN, ROBERT C., Lieut. Col. com'g 64th Ohio, 79, note.

BUCKNER, ALLEN, Colonel 79th Illinois, 79, note. At Franklin, 249. Reporting to Reilly, 250.

BUFORD, ABRAHAM, Brig. Gen. Cavalry Division, 83. At Franklin, 86, 174.

BULLOCK, ROBERT, Colonel com'g Brigade, 132, 135.

BURDICK, JOHN S., Lieut. com'g Battery, 108. Wounded, 187.

BURNSIDE, AMBROSE E., Maj. Gen. Losses at Fredericksburg, 15.

BURR, Frank H. Article cited, 17, note.

CAMPBELL, JOHN A., Major & A. A. G., 39, 57, 60.

CANBY, SAMUEL, Lieut. com'g Battery, 125. In action, ib. Withdrawal, 187.

CAPERS, ELLISON, Colonel 24th South Carolina, 93. Description of field, 93, 94, 154. Wounded, 155. Later career, 155, note.

CAPRON, HORACE, Brig.-Gen. Cavalry, 172.

CARTER, F. B., his house the centre of the line, 43. Topography, 44, 45. Fighting around the house, 97, 116, 119, 142. Experience of his family, 197.

CARTER, JOHN C., Brig. Gen. com'g Brigade, 88. In second line, 93, 132, 152. Attack, 155.

CARTER, M. B. Aids in fixing localities, 43, 44, 77. His narrative, 199. Second line, 242.

CARTER, THEODORIC, Captain. Wounded, 197, 200.

CARTER, Captain 72d Illinois. Effort to retake first line, 161.

CASEMENT, JOHN S. Colonel 103d Ohio, com'g Brigade, 52. Position at Franklin, ib., 112. Attacked, 121, 123. Repulses the enemy, 126, 127. Slight loss, 129, note. Incident, 146, note. Later career, 147, note. Second line, 240. Cox in command on the line, 284, 285.

CASUALTIES at Franklin, compared with other battles, 15. Reilly's brigade, 114. Loring's division, 126, note. Moore's brigade, 134. Bate's division, 135. Chalmers's do., 139, 213. Cockrell's brigade, 213, note. In general, 211 et seq.

CHEATHAM, BENJAMIN F., Maj. Gen. com'g Corps, 17, note. Field headquarters at Franklin, 69. Deployment, 88. Advance, 98, 105. His dead within our works, 118. Lines of attack, 152. Condition of troops, 194.

CHALMERS, JAMES R., Brig. Gen. Cavalry Division, 83. March to Franklin, 84. At Franklin, 86, 89. Line of attack, 133, 135. Report, 138. Casualties, 139.

CLARK, MERVIN, Lieut. Col. 183d Ohio, 57. Killed, 237.

CLAYTON, HENRY D., Maj. Gen. com'g Division, 84. Report quoted, 154. In reserve, 165. Condition of troops, 195.

CLEBURNE, PATRICK R., Maj. Gen. com'g Division, 87, 88. His charge at the centre, 104, 123, 149, 156. Killed, 158. Place, 159. Condition of troops, 194.

CLIFT, DR. His house Schofield's headquarters in morning, 67, 280, note.

COCKERILL, GILES J., Capt. and Chief of Artillery, 49. Guns in

Fort Granger, 59, 95, 123, note. Guns parked, 175, note.
COCKRELL, FRANCIS M., Brig. Gen. com'g Brigade, 88. His attack, 150, 152.
COLLAMORE, G. A., Major, Surgeon. Field work, 181.
COLUMBIA, TENN. The position, 21, 23.
CONRAD, JOSEPH, Colonel 15th Missouri, com'g Brigade. At Winstead Hill, 65. Final outpost position, 73, 76. Asks for directions, 78. His criticism of Wagner, 79, note. The orders, 80. Makes barricade, 92. Hood's centre checked, 103. Message to Wagner, 104. Wedge-shaped line, 131. Groups rallying, 142. Reorganization, 171. Report, 228, 232. Discussion of rallying place, 244 *et seq.* Line of retreat, 245, 246.
COON, DATUS E., Colonel com'g Cavalry Brigade. At Franklin, 172.
COOPER, JOSEPH A., Brig. Gen. Commands detachment at Centerville, 24. Ordered to Franklin, 29. His troops, 54. Expected at Franklin, 173. Devious march to Nashville, 204.
COUGHLAN, JAMES, Lieut. & A. D. C. Sent with order to Opdycke, 95. Killed at the Cotton-Gin, 114.
COX, JACOB D., Brig. Gen. com'g Division, 27. Holds crossing of Duck River, 27, 31. At Spring Hill, 34. March to Franklin, 37. At the Carter House, 39. Ordered to hold Hood back, *ib.* Put in command of Twenty-third Corps, *ib.* Headquarters, 53. Fourth Corps batteries report to him, 57. Schofield's orders for the evening, 68. Wagner's orders, 71. Inspection of lines, 80, 81. Warning to Wagner, 82. Ride to the left, 91. Order to Opdycke, 95. Visits Strickland's line, 142, 144. Visits Ruger, 144. Orders reinforcement from Kimball, 144. Ride to Stiles's brigade, 146. Casement and Reilly, *ib.*, 163. Confidence of holding position, 169. Withdrawal, 186 *et seq.* Orders to Opdycke, 190. Burning building, 191. March to Nashville, 192. Discussion of Wagner's conduct, 220 *et seq.* His visit at Nashville, 222. Friendly judgment, 223. Opdycke's promotion, 225. Letter to Wagner, 226. Preliminary report, 227. Official report, 230, 311. Double breastworks, 233 *et seq.* Wagner's men, 257. Command on the line, 258 *et seq.* Nine points, 260. Schofield quoted, 265. 'Generosity,' *ib.* The 'line' first established, 266. The artillery, 267. Orders to Bridges, 268. To Ruger, 269. To Kimball, 270. Relation to the line, 271 *et seq.*, 291. The Articles of War, 274. Examples of orders, 277, 278. Fighting in the night, 289, note. Report of Twenty-third Corps, 294, 311. Stanley quoted, 296, 297. Personal position, 297. Contemporaneous records, 299, 300. Schofield on the agreed outline, 301, note. Wherry and Opdycke, 301. Method in preparing former histories, 302. Schofield on their accuracy, 303.
COX, THEODORE, Capt. & A. A. G. At Carter House, 53. Wagner's orders, 106. His later career, 107, note. Visits Wagner at river, 146, 225. Sent to Schofield, 169. Stanley wounded and retires, 283. At Schofield's headquarters, *ib.*, 287. Written statements, 336, 338.
CROXTON, JOHN T., Brig. Gen. com'g Cavalry Brigade, 82. At Franklin, 86, 173, 175, 177, 178.
CUNNINGHAM, S. A., Sergt. Major. Description of fight, 164. Death of Strahl, 165.
CUNNINGHAM, W. E., Capt. 41st Tennessee. Description of fight at centre, 164. Wounded, *ib.*

DAVIS, JEFFERSON, visits Hood in Georgia, 2. His approval of Hood's plan of campaign, 3. His speech and its publication, *ib*. Assigns Beauregard to command in Gulf States, 4. Judgment on battle of Franklin, 12.

DAVIS, NEWTON, N., Colonel com'g Brigade. Wounded, 153.

DAWES, EPHRAIM C., Major, 227.

DEAS, ZACHARIAH C., Brig. Gen. com'g Brigade, 88.

DOW, TRIS. T., Major 112th Illinois, Inspector General. On picket at Duck River, 33. Leads reinforcement to Strickland, 146, 161. Sent to Schofield, 170. Covers withdrawal, 186. Scene on the field, 192. Second line, 240. Wagner, 255. Stanley wounded, and retires, 283. Written statement, 332.

DOWLING, PATRICK H., Capt. & Act'g Insp. Gen. Leads a rally, 119. Wounded, 131. Second line, 241.

DUCK RIVER. The position, 23.

ECTOR, MATTHEW D., Brig. Gen. com'g Brigade, 88. Guarding pontoons, *ib.*, 122.

ELLIOTT, WASHINGTON L., Brig. Gen. assigned to Wagner's division, 229.

FATIGUE OF TROOPS. Limits of physical endurance, 203.

FEATHERSTON, WINFIELD S., Brig. Gen. com'g Brigade, 88. His attack, 125, 127, 150.

FIELD FORTIFICATIONS. Problem of attack and defence, 13, 14, 122, 126, 129, 131.

FIELD, H. M., Rev. Cited, 17, note.

FIGUER'S HILL. Topography, 45. Fort Granger *ib.* Artillery there, 48, 94, 125, 175.

FINLEY, JESSE J., Brig. Gen., 88, 132, note.

FIRE DISCIPLINE, 218, 219. Capt. F. N. Maude cited, 219, note.

FLAGS CAPTURED, 110, note, 219, 251, 252, 306.

FORREST, NATHAN B., Maj. Gen. His cavalry corps crosses Duck River, 24, 26. At Spring Hill, 31, 32, 34. At Franklin, 82. The march thither, 83. Covering the infantry, 86. Positions of divisions, 174. Combat at Hughes's Ford, 177. Evening orders, 179. March to Nashville, 196.

FORT GRANGER, 45, 48, 94, 125, 175.

FOSTER, WILLIAM F., Major and Engineer Stewart's Corps, Map, 83; cited, 242.

FRENCH, SAMUEL G., Maj. Gen. com'g Division, 88. Formation and attack, 122, 149.

FRINK, C. S., Major, Surgeon, and Medical Inspector, 40. Schofield's orders, *ib.* Field work, 180 *et seq.* Paper cited, 185, note.

FULLERTON, JOSEPH S., Lieut. Col. & A. A. G., 66. His journal quoted, 67, 70, 229, 271, 288. Wilson's visit to the Truett house, 288.

GENERAL OFFICERS. Fatality amongst, 213, 214.

GIST, W. W. Narrative, 17, note, 201. Private soldier, afterward College Professor, *ib.* Cited, 246, 252.

GIST, S. R., Brig. Gen. com'g Brigade, 88. Advance, 93, 132, 152, 155.

GOODSPEED, WILBUR F., Major of Artillery, 260.

GORDON, GEORGE W., Brig. Gen. com'g Brigade, 88. Advance, 93, 132. Captured, 150, 152. His attack, 155. Quoted, 156. His surrender, 158.

GOVAN, DANIEL C., Brig. Gen. com'g Brigade, 88.

GRANBURY, HIRAM B., Brig. Gen. com'g Brigade, 88. Attack, 156.

GRANT, ULYSSES S., Lieut. Gen. View of the campaign, 4. Losses at Shiloh and Cold Harbor, 15.

Index

GREGG, T. C. Messenger from Conrad to Wagner, 107.
GROSE, WILLIAM, Brig. Gen. com'g Brigade, 61. Position at Franklin, *ib*. Attacked, 133, 136.
GUIBOR'S battery, 84, 89, 126.

HAIGH, Bennett and, History 36th Illinois cited, 241.
HALLECK, HENRY W., Maj. Gen., Chief of Staff of the Army, 10.
HAMMOND, JOHN H., Bt. Brig. Gen. com'g Cavalry Brigade. At Triune, 172; at Wilson's Mill, 173.
HAMPTON, HENRY, Major & act'g A. A. G., 165.
HARRISON, THOMAS J., Colonel com'g Cavalry Brigade at Franklin, 172.
HARTSUFF, WILLIAM, Lieut. Col. & Insp. Gen. Details of withdrawal, 170, 186.
HATCH, EDWARD. Brig. Gen. com'g Cavalry Division, 32. At Franklin, 86, 172. Combat, 177.
HATRY, AUGUST G., Major 183d Ohio, 235. Commands Ruger's skirmishers, 237. Later career, *ib.*, note. Wagner's men, 254.
HAYES, EDWIN L., Lieut. Col. com'g 100th Ohio. Position in line, 59. Fight at centre, 109.
HEARD, J. THEODORE. Major, Surgeon, & Medical Director, 183. Report, 185, note.
HENDERSON, THOMAS J., Colonel 112th Illinois, commanding Brigade, 52. In the fight, 124, note, 146.
HENTIG, FREDERICK G., Capt. & C. S., acting Q. M. Message as to Wagner's men, 145.
HILL, R. J., Major, Surgeon. Description of the Confederate advance, 95. Field work, 183. Field hospital north of river, 286. Dresses Stanley's wound, *ib*.
HOOD, JOHN B., Lieut. Gen. Conference with President Davis, 2. Plan of campaign, 3, 4. Subordinate to Beauregard, 4, 5. Preliminary movements, 5. Delay at Tuscumbia, 6. His forces, 9. Advances from Florence upon Columbia, 11. Criticises his subordinates at Spring Hill, 12. His tactics at Franklin, 14. Turns the position at Duck River, 26. Moves on Spring Hill, 31, 32. Attack repulsed, 33. March to Franklin, 83. His artillery, 84. Determines to attack, 85. Deployment, 87, 88. Field headquarters, 89. Advance of his line, 92. Orders to Cheatham and Stewart, 98. The roar of musketry, 99. His centre retarded, 102. His extreme left, 140. Many lines of attack, 148 *et seq*. Fierceness of fight, 153. His reserves, 165. Chalmers's cavalry, 174. Forrest's, *ib*. Night orders, 194. Condition of troops, *ib*. March to Nashville, 196. Statistics, 207 *et seq*. Recruiting in Tennessee, 208. The night fighting, 289, note.
HOOKER, JOSEPH, Maj.-Gen. Losses at Chancellorsville, 15.
HORN, HENRY, Sergeant com'g Battery, 187.
HORNBROOK, Capt. 65th Indiana, 128, note.
HUNT, WILLIAM W., Capt. 100th Ohio, Acting Major, killed at the parapet, 109.

INGERSOLL, E. J., Capt. 73d Illinois. Quoted, 285.

JACKSON, HENRY R., Brig.-Gen. com'g Brigade, 88. Position in attack. 132.
JACKSON, WILLIAM H., Brig.-Gen. com'g Cavalry Division, 83. At Franklin. 86, 175. Ordered over the Harpeth, *ib*. Combat, 177.
JAMES, WILLIAM, JR., Major 72d Illinois. Wounded, 118, note, 253.
JOHNSON, EDWARD, Maj.-Gen. com'g Division, 31, 84, 88. His attack, 152.
JOHNSON, RICHARD W., Brig.-Gen. com'g Cavalry Division, 82, 172.

Index

JOHNSTON, JOSEPH E., General. Judgment on battle of Franklin, 13.

KIMBALL, NATHAN, Brig.-Gen. com'g Division, 29. At Rutherford's Creek, *ib.*, 32. March to Franklin, 35. Coming into position, 60. His brigades, 61. Orders for the evening, 68. Reinforces Ruger's division, 131. Extent of line, 136. Breastworks not complete, 140. Position firmly held, 145. Withdrawal, 188. Conflict of orders, 189. Relations to Cox, 261, 263. Interpretation of orders, 270.

KINGLAKE'S Invasion of the Crimea, cited, 79, note.

KIRBY, ISAAC M., Brig.-Gen. com'g Brigade. Position at Franklin, 61. Reinforces Moore, 134, note, 136.

LANE, JOHN Q., Colonel 97th Ohio, com'g Brigade. At Winstead Hill, 65. At Privet Knob, 75. Final outpost position, *ib.*, 76. Wagner's orders, 80. Makes barricade, 92. Hood's centre checked, 103. Wedge-shaped formation, 131. Groups rallying, 142. Reorganization, 171. Report, 228, 232. Discussion of rallying place, 244 *et seq.* Line of retreat, 245, 246.

LEE, STEPHEN D., Lieut.-Gen. At Columbia, 31, 33. Follows Schofield's retreat, 34. At Franklin, 84, 152. Report quoted, 153, 154.

LOCUST GROVE. Its place, 55, 245, 246.

LORING, WILLIAM W., Maj.-Gen. com'g Division. Position and advance, 123, 124. Casualties, 126, note. His right, 149.

McCLELLAN, GEORGE B., Maj.-Gen. Losses in Seven-days' battle, 15.

McCoy, DANIEL, Lieut.-Col. com'g 175th Ohio. Helps restore the line, 113.

McDANALD, BEDAN B., Lieut.-Col. com'g 101st Ohio. Reinforces Moore, 134, 234, 264.

McMILLIN, C. W., Major, Surgeon. Field work, 282.

McQUAIDE, JOHN Finding Cleburne's body 59, note.

MANIGAULT, Arthur M., Brig. Gen. com'g Brigade, 88. His attack, 53.

MARCH TO THE SEA. Development of the plan, 7, 8. Number of troops, 8.

MARSHALL, ALEXANDER, Captain of Artillery, 69. Section of battery at Winstead Hill, *ib.*, 95. At Stiles's left, battery in action, 125. Withdrawal, 187.

MAUDE, F. N., Capt. R. E., cited, 219, note.

MAURICE, J. F., Col. R. A. English casualties at Waterloo, 15, note.

MEDICAL DEPARTMENT. Its work, 180 *et seq.*

MERRILL'S HILL. See Privet Knob.

MISSING, THE. Significance of, 216, 247.

MITCHELL, GEORGE H., Lieut. of Artillery. Com'g section with rear guard, 75, 95.

MOHRMANN, Lieut. 72d Illinois. Letter cited, 253.

MOORE, ORLANDO H., Col. 25th Michigan, com'g Brigade, 24, note. Position at Franklin, 54. Repels enemy's attack, 119, 130, 246. Reinforced from Kimball's division, 131, 134. Casualties, 134.

NEELY HOUSE, Hood's headquarters, 89.

OFFICIAL RECORDS. Importance of the publication, 17, 18.

OPDYCKE, EMERSON, Col. 125th Ohio, com'g Brigade. His narrative, 17, note. Rear guard duty, 64. A Winstead Hill 65. In reserve near Carter house, 71. Order from Cox, 95 & note. His charge, 98. In the mêlée, 99, 106. Formation of his troops, 115. At the retrenchment, 142, 145. Second line placed, 170. Withdrawal, 190. Promotion, 225. Difference with Wagner, 224, 231. The second line, 233 *et seq.* Separation from Wagner, 249, 256 Lane's and Conrad's men, 257. Cox's orders, 261, 263. Command on the line, 282. Stanley's presence, *ib.* Stanley wounded and retires, *ib.*

PENNEFATHER, Gen., at Inkerman, 79, note.

PHISTERER, FREDERICK, Capt. U. S. A. Cited, 15, note.

POLK, LUCIUS E., Brig. Gen. com'g Brigade, 88.

PONTOONS, Schofield's request, 24, 39. Arrive but not used, 50, note.

PORTER, PROF. J. G., computes sunset time, 147, note.

POST, P. SIDNEY, Colonel com'g Brigade, 168.

PRATHER, ALLEN W., Colonel 120th Indiana, 125.

PRESSTMAN, STEPHEN W., Colonel of Artillery, 84. Position of guns, 89, 132.

PRISONERS, 211, 215.

PRIVET KNOB. Topography, 46, 73. Cheatham's headquarters, 69, 89. Artillery, *ib.* Lane's skirmishers, 75. Confederate battery, 89.

QUARLES, WILLIAM A., Brig. Gen. com'g Brigade, 88. His attack, 150. Wounded, 151.

RANSBOTTOM, ALFRED, Sergeant 97th Ohio, 251.

REILLY, JAMES W., Brig. Gen. In command of division, 51. Position at Franklin, *ib.* His own brigade, 52. Headquarters, 81. Advance of his second line, 99, 106. Fight at centre, 113. His later career, *ib.*, note. To whom opposed, 150. Second line replaced, 171. Withdrawal, 190. Conrad's men in his line, 250.

Sally in the evening, 251. Wagner's men, 257; Cox in command on the line, 283.

REPEATING RIFLES. A few companies armed with them, 50, note, 111, 126, 127, 217.

REPORTS, value of subordinate ones, 18, 244.

RETRENCHMENT on Carter Hill, 56, 117, 161. The double breastworks at centre, 233 *et seq.*

REYNOLDS, DANIEL H., Brig. Gen. com'g Brigade, 88. Advance, 123.

RICE, HORACE, Colonel com'g Brigade wing, 158. At the works, *ib.*

RODGERS, J. H., Major, Surgeon. Field work, 182.

ROSE, THOMAS E., Colonel 77th Pennsylvania. On skirmish line, 136.

ROSECRANS, WILLIAM S., Maj. Gen. Losses at Stone's River, 15.

ROSS, LAWRENCE S., Brig. Gen. com'g Cavalry Brigade. Combat at Hughes's Ford, 177, note.

ROUSSEAU, LAURENCE H., Lieut. Col. com'g 12th Kentucky. On picket at Duck River, 33. March to Franklin, 35. Position in line, 52, 59. Charge, 99, 111.

RUGER, THOMAS H., Brig. Gen. com'g Division, 24, note. Ordered to Spring Hill, 27, 32, 33. At Thompson's Station, 34. Position at Franklin, 49. His arrival, 53, 54. Artillery, 58, 116. Headquarters, 80. His breastworks, 81, note, 86, 87. Formation of brigades, 118 *et seq.* The attack on Strickland's brigade, *ib.* On Moore's, 130. Reinforced from Kimball's division, 131, 134. Position firm, 145. Withdrawal, 190. Orders to Cooper, 204. Service in Cumberland army, 222. Second line at centre, 239. Understanding of orders, 269.

RUSSELL, EDWARD C., Orderly, carries order to Kimball, 264, note.

RUSSELL, JOHN, Lieut. Colonel,

348 Index

com'g 44th Illinois. In Opdycke's charge, 99, 109, 113, 116.

SANDERS, D. W., Major & A. A. G., 17, note, 44, note. Paper cited, 175, note. Incident, 179, note.

SCHOFIELD, GEORGE W., Lieut. Colonel & Chief of Artillery, 58. Inspects positions, 82. Withdrawal of artillery, 186.

SCHOFIELD, JOHN M., Maj. Gen. Duty assigned to him, 8, 9. Commands in the field under Thomas, 10. Questions of rank with Stanley, ib. Reasons for fighting at Franklin, 11. Ordered to withdraw to Nashville, ib. His qualities as commander, 19, 20. Ordered to delay Hood, 21, 25. The position at Duck River, 26, 27. Asks for pontoon bridge at Franklin, 24. Holding Duck River, 28-32. Thomas's dispatches, ib. At Spring Hill, 32, 33. Reaches Franklin, 37. Communications with Thomas, 38, 40-43. The pontoons, 39. Orders to Cox, 39, 40. Artillery in Fort Granger, 49. Pontoons, 50, note. Orders as to artillery in line, 58. Moves headquarters, 62, 67, 70. Forecast of Hood's action, 68. At Fort Granger, 82, 95, 167. Orders to Wood, ib. Orders from Thomas, 168. Orders midnight withdrawal, 169. Orders to cavalry in morning, 172. Afternoon, 176. Evening, 178, 188. March to Nashville, 192. Statistics, 207 et seq. Preliminary report, 228. Quoted, 265. First orders as to the line, 266. As to artillery, 267. Present in command, 267. Examples of orders, 277, 278. Fighting in the evening, 289, note. The subject in controversy, 301, 303. His report, 305.

SCOFIELD, LEVI T., Capt. & Acting Topographer, 17, note. What he saw at the centre, 104. Wagner's orders, ib. Quoted, 170,

note. Second line, 240. Wagner's men, 255. Stanley wounded and retires, 283. Cox in active command on the line, 291.

SCOTT, THOMAS M., Brig. Gen. com'g Brigade, 88. Attack, 125.

SCOVILL, CHARLES W., Lieut. com'g Battery, 108, 187.

SEARS, CLAUDIUS W., Brig. Gen. com'g Brigade, 88. His attack, 150.

SEDDON, JAMES A., Sec. of War, 212.

SELLON, WILLIAM R., Lieut. Col. com'g Post, 205.

SEXTON, JAMES A., Capt. com'g 72d Illinois, 118, note. Effort to retake first line, 161, 163, note. Later career, ib. Regiment transferred, 210, note. Second line at centre, 239, 253. Wagner's men, 253. General officers on the line, 285.

SHARP, JACOB H., Brig. Gen. com'g Brigade, 88. His attack, 153.

SHAW, THOMAS P., Colonel com'g Brigade. Killed in attack, 153.

SHELLEY, CHARLES M., Brig. Gen. com'g Brigade, 88. Advance, 123, 149. Over our works, 150.

SHERMAN, WILLIAM T., Maj. Gen. Learns Hood's plan, 3. His counter-plan, 4. Correspondence with Grant, ib. His forces, 8. Directions to Thomas, 9, 10. Mentioned, 272, 277, 302.

SHERWOOD, ISAAC R., Lieut. Col. com'g 111th Ohio. Position at Franklin, 131. Repels attack, ib. His missing, 135, note. Second line, 241.

SINCLAIR, WILLIAM H., Major & A. A. G., 67, note, 68.

SMITH, ANDREW J., Maj. Gen. Ordered to reinforce Thomas, 7. Arrives at Nashville, 11, 21, 40, 42, 54, note. Name of his command, 210, note.

SMITH, GEORGE W., Lieut. Colonel com'g 74th and 88th Illinois, con-

Index 349

solidated. In Opdycke's charge, 115.
SMITH, JAMES A., Brig. Gen. com'g Division, 194. Condition of Cleburne's troops, 195.
SMITH, THOMAS B., Brig. Gen. com'g Brigade, 88. Position in attack, 132, 197, 201.
SPARKS, JOSEPH S., Major, Surgeon. Field work, 182.
SPAULDING, OLIVER L., Colonel 23d Michigan. Builds traverses, 81, note.
SPEED, THOMAS, Lieut. & Adjt. 12th Kentucky, 17, note. Paper by him, 43, note, 97, note. His later career, *ib*. Description of fight at the Cotton-Gin, 111.
SPRING HILL. Distance from Columbia, 27, note. Combat there, 33.
STAFFORD, FOUNTAIN E. P., Colonel 31st Tennessee, com'g Brigade, 165.
STANLEY, DAVID S., Maj. Gen. In command of Fourth Corps at Pulaski, 9. Question of rank with Schofield, 10. At Spring Hill, 27, 29, 34. At Franklin, orders to Wagner, 64, 66. At Dr. Clift's, 67. At Truett House, *ib.*, 70. Rides to the front, 98. In the charge, 99. Wounded, *ib*. Retires, 100. At Schofield's quarters, 170. Criticism of Col. Stone's paper, 259. Extent of Cox's command, 259 *et seq*. Nine points, 260, 262. Gratuitous controversy, 265. Assent to essential facts, 267. Misinterpretation of orders, 270. Effect of his appearance at the front, 279 *et seq*. Leaving Schofield, 280. Joins Opdycke, 281. Wherry's statement, 281, note. How long at the front, 282, 283. At the hospital, 286. At Schofield's quarters, 287. Absence from the army and return, 294. Irregular Report, *ib.*, 327. Cox's help, 297. Passing to the left, *ib.* Keeping the field, 297, 298.

STATISTICS of the battle, 207 *et seq*.
STERL, OSCAR W., Colonel 104th Ohio. Report of fight at centre, 112.
STEVENS'S HILL. *See* Winstead Hill.
STEVENSON, CARTER L., Maj. Gen. com'g Division, 84. In reserve, 165.
STEWART, ALEXANDER P., Lieut. Gen. com'g corps, 87. At Franklin, 88, 94. His orders from Hood, 98. Advance of his corps, 122 *et seq.*, 150.
STEWART, ROBERT R., Colonel com'g Cavalry Brigade. At Franklin, 172.
STILES, ISRAEL N., Colonel 63d Indiana com'g Brigade, 52. Position at Franklin, 52, 53, 81, 112. Earliest attacked, 121, 123. The hedge, 124. The railway cut, 125. Repulses the enemy, 128, 129. Sends help to centre, 146.
STOCKTON, JOSEPH, Lieut. Col. com'g 72d Illinois. Wounded, 118, 253.
STONE, HENRY, Capt. & A. A. G., 17, note. Paper in Century War Book, 258. Extent of Cox's command, *ib*.
STONE HILL. *See* Privet Knob.
STOVALL, MARCELLUS A., Brig. Gen. com'g Brigade. Report cited, 166.
STRAHL, OTTO F., Brig. Gen. com'g Brigade, 88. In second line, 93, 132, 152. Attack, 155. Killed, 165.
STREIGHT, ABEL D., Colonel 51st Indiana, com'g Brigade, 168.
STRICKLAND, SILAS A., Col. 50th Ohio. com'g Brigade, 24, note. Position at Franklin, 54. Fight at the centre, 106. Formation of regiments, 118. Reinforced from Stiles, 146, 152, 161. His second line, 233.
SUMAN, ISAAC C. B., Colonel 9th Indiana. Extent of Bate's attack, 137.
SUNSET. Time of, 147, note.

350   Index

TACTICS OF ASSAULT. The lesson of Franklin, 14, 218, 219.

TAYLOR, RICHARD, Lieut. Gen. His department and relations to Beauregard, 4.

TELEGRAPH. Eccentricities of management, 29, 30.

THOMAS, GEORGE H., Maj. Gen. His command, 5, 6, 7. His troops scattered, 6. His problem, 9. Sherman's suggestions, ib. His forces, ib. Orders Schofield to assume command in the field, ib. Intends to command in person when concentration is effected, 10. Urges Schofield to delay Hood, 21, 22. Dispatches to Schofield at Duck River, 21–25. At Franklin, 38, 42, 168. Withdrawing army to Nashville, 192. Orders to Cooper, 204, 205. Calls for reports, 227. Relieves Wagner, 230.

THOMASSON, THEODORE S., Captain com'g Battery, 187.

THOMPSON, CHARLES R., Colonel com'g Brigade, 204, 206.

TRACY, EDWARD E., Capt. & A.D.C. Sent with orders to Opdycke, 96, 261, 263. Dismounted, 262.

TRUETT, ALPHEUS. House Schofield's headquarters, 67.

TURNPIKES at Franklin, 46.

TYLER, ROBERT C., Brig. Gen., 132, note.

TWINING, W. J., Capt. and Chief Engineer Army of Ohio, 27. Reconnoissance up Duck River, ib. His report, 31. At Franklin, 37, 38. Repairing the bridges, 49. His map, 45. Cited, 238, 242, 245.

VAN HORNE, THOMAS B., History of the Army of the Cumberland, 16. Life of Gen. Thomas, 17. Ammunition expended at Franklin, 50, note. Wagner's orders, 74, note. Stanley wounded and leaves the field, 284.

WAGNER, ARTHUR L., Captain U. S. A., cited, 219, note.

WAGNER, GEORGE D., Brig. Gen. com'g Division, 29. At Spring Hill, ib. 35. Rear guard to Franklin, 35. Checking the enemy, 63. Deployed at Winstead Hill, 64. Orders from Stanley, 65, 68. Withdraws to Privet Knob, 69, 70. Places Opdycke in reserve, 71. The division to do the same, 71, 72. Colloquy with Opdycke, 73, 224. New position for Conrad and Lane, 73, 74. Character of the outpost, 77, 78. Orders, 80. At the Carter house, 82. Orders to Conrad and Lane, 103 et seq. Rallying his men, 146. Reorganization, 171. Withdrawal, 188. Conflict of orders, 189. Discussion of his conduct, 220 et seq. Visit to Cox, 222. Opdycke's promotion, 225. His report, 229. Relieved, 230. Rallying place at Franklin, 243 et seq. Line of retreat, 245, 246. Report cited, 248. The order to become the reserve, 261, 272.

WALTHALL, EDWARD C., Maj. Gen. com'g Division, 88. Deployment, ib. Advance, 123, 149. Report, 151.

WHEELER, JOSEPH, Maj. Gen. His information to Hood, 5.

WHERRY, WILLIAM M., Major & A. D. C. 82, 169. Letter giving incidents, 280, note. Later career, 281, note. Stanley at Schofield's headquarters, ib. Cox commanding on the line, 291.

WHITAKER, WALTER C., Brig. Gen. com'g Brigade. At Spring Hill, 34. At Franklin, 62, 136.

WHITE, JOHN S., Lieut. Col. com'g 16th Kentucky. On picket at Duck River, 33, note. Position at Franklin, 52, 59. Charge, 99, 109. Wounded, 114.

WHITE, LYMAN A., Lieut. com'g Battery, 187.

WHITESIDES, EDWARD G., Capt., Act'g A. A. G. Bearer of message, 69.

WILSON, JAMES H., Bt. Maj. Gen. Commands cavalry corps, 10. Resists Forrest at Duck River, 24, 26, 27. To keep touch with infantry, 32. Connection broken, 33. Joins Schofield at Franklin, 41. Cavalry dispositions, 82, 168, 172 *et seq.* Headquarters, 172. Hammond's brigade at Triune, *ib.* Advance at Hughes's Ford, 177. Engagement there, *ib.*, *et seq.* At Schofield's quarters, 178. Orders as to withdrawal, 178. March to Nashville, 196. Stanley at Schofield's quarters, 287. The hour, 288.

WINSTEAD HILL. Topography, 46. Wagner halts there, 64, 66, 68, 71.

WOOD, THOMAS J., Brig. Gen. com'g division, 26. March to Franklin, 35. Position north of Harpeth River, 62. Ordered to cover crossing of Harpeth, 167. To send brigade to Hughes's Ford, 168. Disposition of troops, *ib.* To cover withdrawal, 178. Report, 228, note, 323. Commanding Fourth Corps, 293.

ZIEGLER, JACOB, Captain of Artillery. Position at Franklin, 117, note, 131, 133. Withdrawal, 187.